Design Thinking

Design Thinking is a set of strategic and creative processes and principles used in the planning and creation of products and solutions to human-centered design problems.

With design and innovation being two key driving principles, this series focuses on, but is not limited to, the following areas and topics:

- User Interface (UI) and User Experience (UX) Design
- Psychology of Design
- Human-Computer Interaction (HCI)
- Ergonomic Design
- Product Development and Management
- Virtual and Mixed Reality (VR/XR)
- User-Centered Built Environments and Smart Homes
- Accessibility, Sustainability and Environmental Design
- Learning and Instructional Design
- Strategy and best practices

This series publishes books aimed at designers, developers, storytellers and problem-solvers in industry to help them understand current developments and best practices at the cutting edge of creativity, to invent new paradigms and solutions, and challenge Creatives to push boundaries to design bigger and better than before.

More information about this series at `https://link.springer.com/bookseries/15933`.

The Persuasive Agent

UX Design Psychology for Agentic AI

Victor Yocco

Apress®

The Persuasive Agent: UX Design Psychology for Agentic AI

Victor Yocco
Glenside, PA, USA

ISBN-13 (pbk): 979-8-8688-3316-8 ISBN-13 (electronic): 979-8-8688-3317-5
https://doi.org/10.1007/979-8-8688-3317-5

This book contains AI generated images generated by the author.

Managing Director, Apress Media LLC: Welmoed Spahr
Acquisitions Editor: James Robinson-Prior
Editorial Assistant: Gryffin Winkler

Cover designed by eStudioCalamar

Distributed to the book trade worldwide by Springer Science+Business Media New York, 1 New York Plaza, New York, NY 10004. Phone 1-800-SPRINGER, fax (201) 348-4505, e-mail orders-ny@springer-sbm.com, or visit www.springeronline.com. Apress Media, LLC is a Delaware LLC and the sole member (owner) is Springer Science + Business Media Finance Inc (SSBM Finance Inc). SSBM Finance Inc is a **Delaware** corporation.

For information on translations, please e-mail booktranslations@springernature.com; for reprint, paperback, or audio rights, please e-mail bookpermissions@springernature.com.

Apress titles may be purchased in bulk for academic, corporate, or promotional use. eBook versions and licenses are also available for most titles. For more information, reference our Print and eBook Bulk Sales web page at http://www.apress.com/bulk-sales.

Any source code or other supplementary material referenced by the author in this book is available to readers on GitHub. For more detailed information, please visit https://www.apress.com/gp/services/source-code.

If disposing of this product, please recycle the paper

To Max, Emily, and Ella, you inspire me daily.

Table of Contents

About the Author xv

About the Technical Reviewer xvii

Acknowledgments xix

Introduction xxi

Part I: Foundations of Agentic Persuasion 1

Chapter 1: The Trust Threshold: From Guiding Clicks to Acting on Behalf 3

1.1 What We Talk About When We Talk About Persuasion 4

1.2 The Persuasion Gap: From Interface to Agent 6

1.3 The High Cost of Getting Trust Wrong 9

1.4 A Roadmap for Designing Trustworthy Agents 12

Part 1: Foundations of Agentic Persuasion (Chapters 1–4) 12

Part 2: The Psychological Toolkit for Trust (Chapters 5–9) 13

Part 3: Ethical Persuasion in Practice (Chapters 10–12) 13

1.5 Summary 14

1.6 Chapter 1 Workshop: The Trust Threshold Audit 15

Part 1: Auditing a Product You Use Every Day 15

Part 2: Auditing Your Own Product (15 minutes) 16

Part 3: Imagining the Agentic Leap (20 minutes) 17

1.7 References and Further Reading 19

Chapter 2: A Brief History of Digital Influence....................................23

2.1 B.J. Fogg and Captology: The Three Ingredients of Behavior........................23

2.2 The Hook Model: Building Habit-Forming Products 28

2.3 Why These Models Are No Longer Enough...33

2.4 Adapting the Foundations for an Autonomous World.....................................35

Motivation Becomes Motivating Delegation 37

Ability Becomes Simplicity of Trust 37

The Reward Becomes the Competence Loop...38

Investment Becomes Teaching and Correction 39

2.5 Summary...40

2.6 Chapter 2 Workshop: Deconstructing Influence..41

2.7 Your Turn: A Guided Analysis 43

2.8 References and Further Reading 45

Chapter 3: Debugging the Conversation: A Communication Framework for Agent Design...49

3.1 Beyond Intuitive: A Formal Model for Interaction..50

3.2 Deconstructing Conversation: Senders, Receivers, Messages, and Channels 52

Senders and Receivers...53

Message 53

Encoding and Decoding..54

Channel 54

Feedback..55

Field of Experience..55

3.3 Finding the Noise: Technical, Semantic, and Psychological Interference......56

Technical Noise 57

Semantic Noise 58

Psychological Noise..59

3.4 Using the Model to Diagnose a Failed AI Interaction....61
3.5 Summary....64
3.6 Chapter 3 Workshop: The Communication Autopsy....66
3.7 References and Further Reading....70

Chapter 4: The Agent As Social Actor....75

4.1 Why We Are Polite to Computers....76
4.2 Unconscious Social Heuristics in the Age of AI....80
4.3 The Agent's Persona Is Not Optional....83
4.4 Opportunities and Risks of Being a Social Actor....84
Opportunities: Leveraging Social Heuristics for Good....85
Risks: The Uncanny Valley of Social Design....86
4.5 Summary....88
4.6 Chapter 4 Workshop: The Social Actor Audit....89
Part 1: The CASA Safari....89
Part 2: The Accidental Persona Audit....90
Part 3: The Persona First Sketch....92
4.7 References and Further Reading....93

Part II: The Psychological Toolkit for Trust....97

Chapter 5: First Impressions: Onboarding and Designing the Agent's Persona....99

5.1 You Only Get One Introduction....100
5.2 Persona Canvas: Defining Your Agent's Identity....102
Core Archetype and Role....104
Key Personality Traits (Defined on a Spectrum)....105
Communication Style (Voice and Tone)....107
Domain Expertise and Boundaries....108
Error Philosophy....109

5.3 Onboarding for Trust: Setting Expectations and Demonstrating Value........109

Step 1: Establish the Persona........110

Step 2: Set Clear Expectations........111

Step 3: Deliver an Immediate Win........112

5.4 The Agent's Body Language: How Motion, Timing, and Micro-interactions Express Personality........113

5.5 Summary........116

5.6 Chapter 5 Workshop: The Blueprint for a First Impression........117

Part 1: The Persona Deconstruction........117

Part 2: Your Turn: The Agent Persona Canvas........118

Part 3: Scripting the First 30 Seconds........120

5.7 References and Further Reading........121

Chapter 6: Foundations of Influence: Reciprocity, Liking, and Authority........125

6.1 Reciprocity: The Power of Giving Value First........127

Reciprocity in the Interface Era........128

6.2 Liking: Building Rapport Through Similarity and Praise........131

Similarity: The "You're Like Me" Heuristic........132

Praise: The Power of Authentic Compliments........133

Cooperation: The "We're in This Together" Frame........134

6.3 Authority: Communicating Expertise and Credibility........135

Pattern: The Expert Introduction........135

Pattern: The Confident Expert Who Admits Ignorance........136

Pattern: Showing Your Work (Borrowed Authority)........137

6.4 Ethical Checkpoint: The Dark Side of Influence........139

The Ethical Litmus Test........140

6.5 Summary........141

6.6 Chapter 6 Workshop: Designing the Relationship 142
Part 1: The Influence Audit 142
Part 2: The Relationship Blueprint 143
Part 3: The Ethical Checkpoint 145
6.7 References and Further Reading 146

Chapter 7: Mechanics of Influence: Social Proof, Scarcity, and Consistency 151

7.1 Social Proof: Using the Wisdom of the Crowd Responsibly 152
Social Proof in the Interface Era 153
7.2 Scarcity: The Difference Between Genuine Urgency and a Dark Pattern 156
The Ethical Litmus Test for Scarcity 157
7.3 Consistency: The Foot-in-the-Door Technique for Building Delegation 160
The Agentic Upgrade: The Delegation Ladder 160
7.4 Ethical Checkpoint: When Social Proof Creates Bias 164
The Ethical Litmus Test for Social Proof 165
7.5 Summary 166
7.6 Chapter 7 Workshop: The Ladder of Trust 167
Part 1: The Scarcity and Social Proof Audit 168
Part 2: Designing Your Agent's Delegation Ladder 169
Part 3: The Socially Aware Action Enabler 171
7.7 References and Further Reading 172

Chapter 8: The Transparency Paradox: Explainability and Calibrated Trust 177

8.1 Surprising Research on AI Disclosure 178
8.2 The Goal Is Calibration, Not Absolute Trust 180
8.3 Transparency Triage Framework: What, When, and How to Explain 182
Quadrant 1: Low Risk and Expected (The "Just Do It" Zone) 183
Quadrant 2: Low Risk and Unexpected (The Huh? Why? Zone) 184

Quadrant 3: High Risk and Expected (The Just Confirm Zone) 185
Quadrant 4: High Risk and Unexpected (The Whoa, Hold On! Zone) 186
8.4 Designing for Explainability on Demand 187
Pattern: Progressive Disclosure: The Expandable Why? Link 187
Pattern: The Interactive Slider or Control 189
Pattern: The "Show Your Work" Transcript 190
8.5 Summary 193
8.6 Chapter 8 Workshop: Triage and Transparency Blueprint 194
Part 1: Triage Drill 194
Part 2: Designing for the "Whoa, Hold On!" Moment 196
Part 3: The Transparency Blueprint for Your Agent 199
8.7 References and Further Reading 201

Chapter 9: Navigating Cognitive Bias: Minds, Models, and Mental Shortcuts 205

9.1 Our Brains on Autopilot: A Cognitive Bias Primer 206
9.2 Automation Bias: Over-trusting the Machine 208
9.3 Confirmation Bias: Seeing What We Want to See in AI Outputs 210
9.4 Designing Interventions to Counteract Common Biases 212
Interventions for Automation Bias 213
Interventions for Confirmation Bias 216
9.5 Matching the Intervention to the Persona 220
9.6 Summary 222
9.7 Chapter 9 Workshop: Architecting the Cognitive Environment 223
Part 1: The Bias Spotter 224
Part 2: Designing Your Cognitive Speed Bumps 225
Part 3: The Persona Alignment Check 227
9.8 References and Further Reading 228

Part III: Ethical Persuasion in Practice 233

Chapter 10: The Guardian Agent: Designing for the User's Stated Goals 235

10.1 From "Nudge" to "Guardian" 236

10.2 Pattern: Introducing Positive Friction to Prevent Errors 238

The "Are You Sure?" Supercharged 239

10.3 Pattern: The Cooling-Off Period 239

10.4 Pattern: Using Reflection Prompts to Align with Long-Term Goals 242

The Anatomy of a Good Reflection Prompt 242

The Digital Well-Being Guardian 243

The Financial Guardian 244

The Learning Guardian 244

10.5 The Non-negotiable Off Switch: Ensuring User Control 245

Layer 1: The Snooze/The Override 246

Layer 2: The Granular Settings 246

Layer 3: The Nuclear Option: The Hard Off Switch 247

10.6 Summary 249

10.7 Chapter 10 Workshop: Building Your Guardian 250

Part 1: The Guardian Audit of Your Own Life 251

Part 2: Designing Your Guardian's Interventions 252

Part 3: The Blueprint for Autonomy (The "Off Switch") 254

10.8 References and Further Reading 256

Chapter 11: Measuring Trust: Research Methods for Agentic Systems 261

11.1 You Can't Improve What You Don't Measure 262

11.2 Quantitative Methods: The Delegation Test and Trust Scales 265

The Delegation Test: Measuring Trust Through Behavior 265

Trust Scales: Measuring Trust Through Attitude 269

A Sample Nine-Item Trust Scale 270

11.3 Qualitative Methods: The "Wizard of Oz" Study for Agents 271

How to Run a Wizard of Oz Study 272

11.4 The Diary Study: Tracking Trust Over Time 275

Example Diary Study: Understanding Trust in a New AI-Powered Financial Advisor 276

11.5 Summary 280

11.6 Chapter 11 Workshop: The Trust Measurement Plan 283

Part 1: Your Quantitative Toolkit 283

Part 2: Your Qualitative Toolkit 286

11.7 References and Further Reading 288

Chapter 12: Persuasion in the Wild: Case Studies and Future Directions 293

12.1 Deconstructing a Shopping Agent 294

Case Study 1: ThreadFinder, the AI Personal Stylist 294

12.2 Deconstructing a Productivity Copilot 297

Case Study 2: Notesy, the AI Meeting Assistant 297

12.3 Deconstructing a Travel Planner 301

Case Study 3: Wanderlust AI, the Vacation Planner 301

12.4 Deconstructing a Coding Copilot 305

Case Study 4: Forge, the AI Coding Copilot 305

12.5 The Future of the Persuasive Agent: Your Role in Shaping Ethical AI 312

12.6 Summary 313

12.7 The Final Workshop: Grade the Agent 314

The Case: Sprout, the First-Job Money Coach 314

12.8 References and Further Reading 319

Index 321

About the Author

Victor Yocco, PhD, is a UX research leader and author whose work focuses on design psychology and communication. For two decades, he has been at the forefront of user experience research, leading teams that bridge the gap between human behavior and product design. His book, *Design for the Mind* (Manning), successfully translated complex psychological principles into actionable guidance for tech professionals. His recent work has been dedicated to researching the unique UX challenges of generative and agentic AI with articles being published in *Smashing Magazine* and research journals on designing for trust and explainability in AI, and a monthly column on psychology and AI in *UXMatters*.

About the Technical Reviewer

Angelica Lo Duca is a researcher at the Institute of Informatics and Telematics of the National Research Council, Italy. Her research interests include data storytelling and AI-based decision support applications, with interests in data visualization and applied AI for education, cultural heritage, tourism, and environmental monitoring. She is the author of *Comet for Data Science* (Packt, 2022), a co-author of *Learning and Operating Presto* (O'Reilly, 2023), and the author of *Data Storytelling with Altair and AI* (Manning, 2024), *Become a Great Data Storyteller* (Wiley, 2025), and *Learning Generative AI Tools for Excel* (O'Reilly, 2026).

Acknowledgments

A book is rarely the product of a single mind, and this work is no exception. I owe a debt of gratitude to all the remarkable researchers and designers I have had the privilege of working alongside over the years. Your insights, rigorous debates, and shared dedication to human-centered technology laid the intellectual foundation for the frameworks in these pages.

I am deeply grateful to the entire team at Apress for turning this manuscript into a finished reality:

James Robinson-Prior, Acquisitions Editor, for seeing the immediate value in my original proposal and championing this project from day one. Gryffin Winkler, for your superb communication and guidance. Sowmya Thodur and Sriram N, Production Editor, for your diligence and relentless work in keeping the project moving smoothly forward. Angelica Lo Duca, Technical Editor and dear friend, whose sharp feedback sharpened my arguments and whose friendship made the drafting journey a joy.

Crossing the finish line of a book requires a strong support system, and I am fortunate to have the very best: Eleanor Yocco, for your endless love, warm hugs, funny jokes, and the confidence you gave me to keep moving forward. Emily Yocco, for always cheering me on and sending steady positive thoughts my way. Max Yocco, for your constant support, encouragement, and insightful conversations. Amy Dalsan, for supporting me throughout this journey and keeping me motivated whenever the finish line felt far away. Splinta Claus, for always finding a way to make me smile when I needed a break.

Finally, a special nod to Glitch, our faithful AI companion featured throughout this book, thank you for serving as a constant spark of visual and creative inspiration.

Introduction

Something happens around the third or fourth chapter of a book like this one. You meet a framework, you like it, you file it away. The next chapter hands you another framework, and the one before it starts to fade. By the halfway mark you carry a drawer full of tidy models, each sound on its own, none of them obviously wired to the others. A careful reader of my drafts named that exact feeling, and she was right to. So before you meet the first framework, I want to hand you the single idea that holds all of them together.

That idea arrives in Chapter 1, and it runs under every page that follows. When a person stops operating software and starts handing tasks to it, the trust they need changes character. Clicking a button asks for Usability Trust, the plain belief that a tool will do what its label promises. Telling an agent to book the flight, move the deadline, or spend the money asks for Delegation Trust, the belief that a competent partner will make a good call on your behalf. Between those two kinds of trust sits a gap I call the Trust Threshold. Users stand at its edge and hesitate. Your work as a designer is to build the crossing.

In Chapter 1, you will read about an e-commerce personal shopper that a team built over a year. The recommendation engine was strong. In tests, it picked clothes people genuinely liked. The engine worked. Users left anyway, inside a month, because the product asked them to hand over their taste to a stranger on the first date. The failure was trust, not code.

Not every failure looks like slow abandonment. Some arrive in the first five seconds. In Chapter 5 I describe a budgeting app that greets users with breezy slang about my money before it shows a single screen of value. I'd delete it on the spot. One product lost users over a month, the other over five seconds, and both stumbled at the same place: the threshold. Every pillar in this book exists to keep you from that fall.

Picture the crossing as a bridge. Every framework in these chapters is a pillar under its deck, and each pillar carries a specific part of the load. You do not have to memorize nine separate frameworks. You learn one bridge and know which pillar to reach for when a real problem lands on your desk. Here they are, in the order you meet them, which is also a workable order for building an agent from scratch.

The Delegation Gap (Chapters 1 and 2). Turn here when adoption stalls at the handoff and you need to understand why a user will not let go of a task.

Communication Diagnostics (Chapter 3). This is your fault locator. When an interaction feels wrong and nobody can say why, the Transactional Model and its three kinds of noise turn a vague complaint into a located fault.

The Social Actor Lens (Chapter 4). Use it when your agent is technically right and still disliked. People treat it as a person, so it lives by social rules.

Persona and First Impression (Chapter 5). When the agent's voice wanders, start here. The Persona Canvas and the onboarding sequence decide who is talking before the first task begins.

The Influence Toolkit (Chapters 6 and 7). This pillar earns the yes. When you need the user to act and want to do it mindfully, Cialdini's principles become concrete design patterns.

Transparency (Chapter 8). Reach for it when users either swallow every answer or doubt all of them. This pillar sets how much reasoning to show, and when.

The Cognitive Environment (Chapter 9). Come here to guard the user's own judgment against bias, echo chambers, and thoughtless acceptance. The Cognitive Speed Bump slows the moments that warrant slowing.

The Guardian Agent (Chapter 10). Keep this one running at all times. It checks that every persuasive choice serves the user and not the house, and it is the conscience of the system.

Measurement and Evidence (Chapters 11 and 12). Reach for it to find out whether the trust you designed is real. The research methods measure it, and the case studies grade finished agents against the whole bridge.

Read the chapters in order the first time through. The pillars build on each other, and each one assumes the ground laid by the last. After that first pass, treat the book as a bridge you can walk to any point. When something breaks in your own product, name the symptom, find the pillar, and pull the framework that sits there. A user who will not delegate is a pillar 1 problem. An agent that feels cold is a pillar 3 or 4 problem. A recommendation nobody acts on is a pillar 5 problem. The diagnosis leads you straight to the tool.

Figure 1 represents the whole crossing we will cover in this book. Come back to this map whenever the chapters start to feel like separate rooms. They are not meant to stay separate. They exist best as supporting pillars, and this is the bridge they hold up.

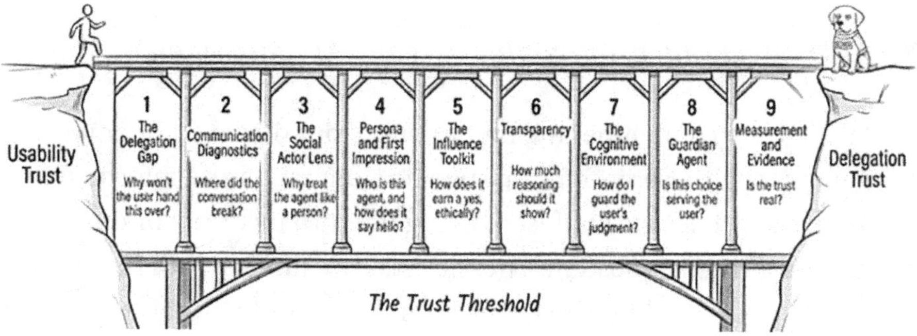

Figure 1. *The Trust Bridge. We explore nine pillars in this book to learn how to help users cross the trust threshold*

Glitch, Our Faithful AI Companion

Meet Glitch, our faithful dog companion shown in Figure 2. You'll see Glitch in figures and examples throughout our book. I've created Glitch as a visual anchor. He sits alongside our agentic AI diagrams to represent the everyday realities of artificial intelligence. We designed AI to be friendly and ready to serve us as a loyal partner. A well-trained model might draft an email to a client or summarize notes from a visit to a local cafe.

Figure 2. *Meet Glitch, our faithful agentic AI companion*

We must also remember the limitations of our digital companions. Developers train these systems on specific commands and information. An AI can easily become confused. It might return incorrect information when asked a simple question. These mistakes cause users to hesitate at the Trust Threshold. When an agent stumbles, users lose their Delegation Trust and abandon the task. Glitch reminds us to build a strong trust bridge for the ideal outcome while preparing for these inevitable errors.

PART I

Foundations of Agentic Persuasion

CHAPTER 1

The Trust Threshold: From Guiding Clicks to Acting on Behalf

Welcome to the new paradigm of software development. If you are reading this, you are likely standing at the edge of a significant transition in your career. You have spent years mastering the craft of building interfaces—screens, buttons, forms, and flows that respond to user input with predictable precision. You know how to minimize cognitive load and guide a user from point A to point B.

But now, the requirements have changed. You are no longer being asked to build tools that wait for a command. You are being asked to build agents that think, decide, and act.

A quick grounding on terms we will use throughout the book: an AI agent is a software system that does not just respond to your commands. It perceives context, forms a plan, and carries out multistep tasks autonomously on your behalf. Think of the difference between a search engine that displays flight options, a tool you operate, and a digital assistant that researches, compares, and books the best flight while you are in a meeting, an agent that acts for you. Agentic AI is the broader

V. Yocco, *The Persuasive Agent*, Design Thinking,
https://doi.org/10.1007/979-8-8688-3317-5_1

paradigm: systems designed for this kind of goal-directed, autonomous action. Throughout this book, "agent" and "agentic AI" refer specifically to this class of system, not tools that wait, but actors that do.

This transition from interface design to agent design is not merely a technical upgrade. It represents a fundamental change in the psychological contract between the user and the software. When we build standard applications, we ask users to trust the *utility* of the tool. When we build Agentic AI, we ask them to trust the *judgment* of the machine.

This chapter defines that difference. We will explore the concept of the Trust Threshold—the psychological gap that users must cross before they are willing to delegate tasks to an AI. We will examine why well-engineered agents fail when they ignore this threshold, and we will lay the groundwork for a new approach to product development that treats trust as a primary architectural component.

1.1 What We Talk About When We Talk About Persuasion

Let's clear the air right away. For many people in the tech world, the word "persuasion" lands with a thud. It feels vaguely sinister, like something you'd discuss in a windowless boardroom while stroking a white cat. It brings to mind dark patterns, manipulative pop-ups, and the general sense that someone is trying to pull a fast one. If you just recoiled a little bit, that's a perfectly healthy reaction. It means your ethical compass is still pointing north.

But persuasion, at its core, is a fundamental human act. It is the process of communication that leads to a voluntary change in beliefs, attitudes, or behavior. When you recommend a restaurant to a friend, you are persuading. When a fitness app encourages you to take the stairs instead of the elevator, it is persuading. When a well-designed button with a clear label invites a user to click "Save," it is, in its own small way, persuading.

In the context of design, ethical persuasion is the art and science of guiding users toward actions that are in their own best interest and align with their stated goals. It's not about trickery; it's about clarity, motivation, and the removal of psychological friction. Manipulation, its villainous cousin, is the use of deception or coercion to guide users toward actions that serve the system's goals, often at the user's expense. The techniques can sometimes look similar on the surface, but the intent is what separates a helpful guide from a con artist. A chef's knife can be used to prepare a life-sustaining meal or to cause harm. The morality lies not in the tool, but in the intent of the person wielding it. This book is a course in ethical culinary skills.

For as long as we have been designing digital experiences, we have been in the business of persuasion. We have been persuading users to trust that our websites are secure enough for their credit card information. We have persuaded them that signing up for our newsletter will provide genuine value and not just spam. We have used principles of visual hierarchy, social proof (customer reviews!), and clear copywriting to persuade them that clicking a particular button is the right next step to accomplish their goal.

For the last two decades, this has been the landscape. The persuasion was gentle, nudge-oriented, and always left the user in the driver's seat. We were persuasive co-pilots, pointing out the best route on the map. But the map is changing. In fact, the map is being replaced entirely, and our role along with it. We are no longer just co-pilots. We are being asked to build the chauffeur. And that changes absolutely everything.

1.2 The Persuasion Gap: From Interface to Agent

A decade ago, the central challenge of UX was usability. The digital world was messy. The goal was to equip teams with cognitive psychology basics to build cleaner, simpler interfaces. We fought a war against cognitive load, striving for effortless usability where users did not have to think about the UI.

The core assumption of that time was that the user remained in direct control. Every significant action resulted from a conscious, physical manipulation: a click, a tap, or a keypress.

The user placed their trust in the predictable mechanics of the interface. I call this **Usability Trust**. It's the belief that the tool will function as expected. When I click the "B" icon in a word processor, I have a high degree of Usability Trust that it will make my text bold. I don't need a personal relationship with the icon. I don't need it to understand my goals. I just need it to work. For 20 years, building Usability Trust was the primary job of a UX professional.

Then, everything changed. The slow, steady evolution of digital assistants, recommendation engines, and chatbots suddenly accelerated into the Cambrian explosion of generative AI. We have entered a new design paradigm. The products we build now accept ambiguous goals. They synthesize novel insights and perform multistep actions on our behalf. This creates a massive chasm in the user experience landscape; a gap between what we used to ask of users and what we are asking now.

We have gone from designing interfaces to designing agents. And an agent requires a profoundly different and deeper form of trust. Consider a flight-booking example to illustrate this gap, as shown in Figure 1-1.

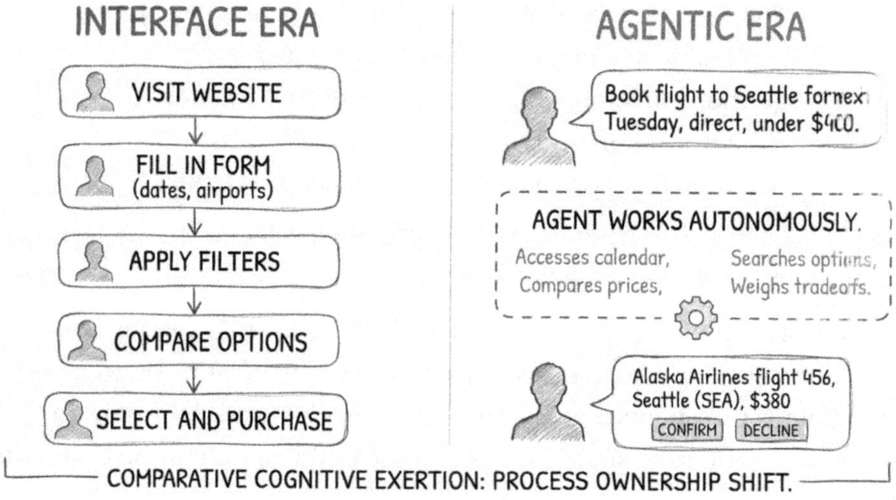

Figure 1-1. *The Two Eras of Digital Interaction. Image generated by author using Gemini Pro 3.1/Nano Banana 2*

The Interface Era (Usability Trust): You visit a website. You see a form with fields for your departure city, destination, and dates. You see a big, friendly calendar widget. You choose your options, apply some filters, and the system presents you with a list of flights. You, the user, perform the cognitive work of comparing the options, weighing the trade-offs between price and layover time, and making the final selection. When you find the one you want, you click "Select" and "Purchase." At every step, you are in direct control. Your trust is placed in the interface: "I trust this form will correctly capture my dates" and "I trust this button will book the flight I selected."

> **The Agentic Era (Delegation Trust):** You open an app and say, or type, "Find me a good flight to our quarterly meeting in Denver next month. You know the dates are on my Outlook Calendar. Prioritize a non-stop flight on the way there so I can make the morning session, but I don't mind a layover on the way back to save some money."

The agent goes to work. It doesn't show you a form. It already knows your home airport, it accesses your calendar for the meeting dates, and it understands the nuanced goal—speed on the outbound, economy on the return. It performs the search, compares the options, weighs the trade-offs, and presents you with a recommendation: "I've found a direct flight on United arriving at 9:10 AM, and a return flight on Southwest with a one-hour layover in Chicago that saves you $120. The total is $435. Shall I book it using your corporate card?"

This is a fundamentally different kind of interaction. The agent has performed cognitive work that you used to do. It has made judgments and decisions on your behalf. To agree to this suggestion, you need more than Usability Trust. You need **Delegation Trust**. Delegation Trust is the belief that the agent understands your intent, is competent to execute the task, and is aligned with your best interests. It is the kind of trust you place in a human assistant. It is personal, relational, and built on a foundation of perceived competence and shared goals.

The gap between Usability Trust and Delegation Trust is what I call the **Trust Threshold**, as shown in Figure 1-2. The term Trust Threshold is my own framing, developed through a review of the existing literature on automation trust and human-AI reliance. It is a psychological chasm, and asking a user to cross it is the single greatest challenge in modern product design. You cannot simply build a technically proficient agent and expect users to leap across this chasm. They won't. They will stand on the edge, look at the other side with suspicion, and retreat to the familiar comfort of the old interface. Our job, as the creators of these new systems, is to build a bridge. This book provides the psychological blueprints for that bridge.

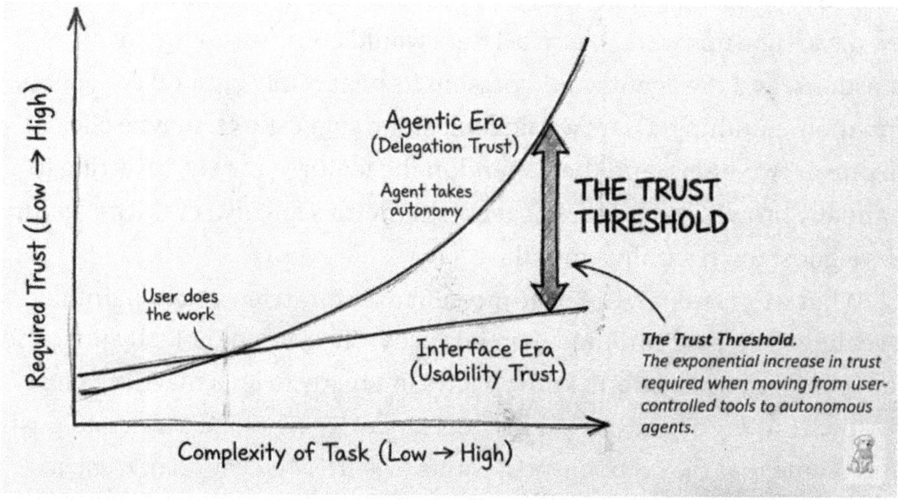

Figure 1-2. *The Trust Threshold. Image generated by author using Gemini Pro 3.1/Nano Banana 2*

1.3 The High Cost of Getting Trust Wrong

Building a bridge of trust is not a "nice-to-have" or a soft UX metric. In the agentic era, it is a direct prerequisite for product viability. When a traditional website has poor usability, the cost is friction. Users get frustrated, tasks take longer, and some people might give up. But when an agent fails to earn trust, the cost is outright rejection.

I was once brought in to consult on a project for a large e-commerce company. They had built a brilliant, technically sophisticated AI-powered "personal shopper." The idea was that you would tell it your style preferences, your budget, and what kind of event you were shopping for, and it would automatically curate a collection of outfits for you. The engineering team had spent over a year on the recommendation engine, and it was remarkably good. During internal tests, it consistently picked items that users genuinely liked.

The product launched with a huge marketing push. For the first few days, sign-ups were massive. Users would come in, answer a few questions, and the agent would present its beautifully curated collection. And then... nothing. Users would look at the suggestions, maybe click on one or two items, and then abandon the feature entirely, reverting to manually browsing the site. Engagement metrics fell off a cliff. The feature was a ghost town within a month.

What went wrong? The post-mortem user interviews were painfully revealing. One user summed it up perfectly: “It seemed perfectly fine. It felt like it knew me, but I don’t know *it*. I wasn’t ready to let a machine I just met pick out my clothes. What if it gets it wrong? What if it’s trying to push stuff on me that the company just wants to sell?” Delegation of trust, as shown in Figure 1-3, was the failure point.

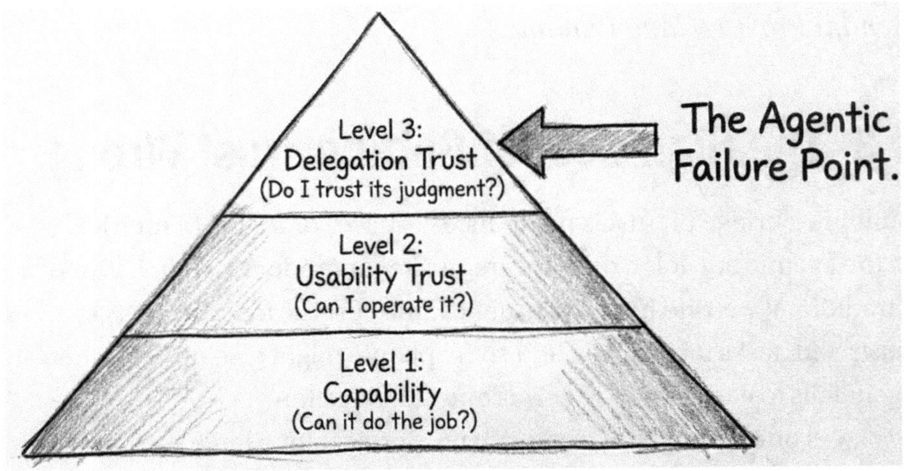

***Figure 1-3.** The agentic failure point is often the point where the user must delegate trust. Image generated by author using Gemini Pro 3.1/ Nano Banana 2*

The product was technically brilliant but psychologically naive. It had asked for Delegation Trust on the first date. It tried to leap across the Trust Threshold without even introducing itself properly. The company had invested millions of dollars and thousands of engineering hours into a feature that failed not because it didn't work, but because it didn't do the human-centric work of building a relationship first. The cost of getting trust wrong was a complete write-off of a massive investment.

This story is not an outlier. It is the default outcome for teams that approach agentic AI as a purely technical problem. We see it everywhere:

- **The Overeager Assistant:** Smart home devices that proactively offer suggestions that feel more intrusive than helpful, leading users to disable the very features that are meant to be the core value proposition.
- **The Black Box Recommender:** Media platforms whose algorithms serve up strange or inappropriate content without explanation, eroding the user's confidence in the system's judgment.
- **The Unreliable Copilot:** AI coding assistants that generate plausible-looking but subtly incorrect code, destroying a developer's trust and forcing them to spend more time verifying the output than it would have taken to write the code themselves.

In each case, the failure is a failure of trust. The cost is measured in abandoned features, uninstalled apps, damaged brand reputation, and staggering amounts of wasted development resources. Getting trust wrong is not a UX problem. It is a business catastrophe.

1.4 A Roadmap for Designing Trustworthy Agents

If the challenge is this great and the stakes are this high, where do we even begin? The solution is not to simply add a friendlier avatar or a witty loading message. "UX theater" without a solid foundation will be exposed as hollow almost immediately. The solution is to fundamentally change how we approach the design process. We must become applied psychologists. This is user-centered design carried to its next stage: we still start from the person's goals, but now we design a collaborator that acts on them rather than a screen that waits for them.

This book provides a roadmap for doing just that. It is a structured, systematic approach to building the bridge across the Trust Threshold. Our journey will be organized into three parts, each building on the last.

Part 1: Foundations of Agentic Persuasion (Chapters 1–4)

Before we can build, we need to understand our materials. In this first part, we will establish the core concepts that underpin the rest of the book. We will define the Trust Threshold, explore the history of digital influence to understand where our ideas have come from, and introduce two powerful mental models: the Transactional Model of Communication for diagnosing interaction failures, and the CASA (Computers Are Social Actors) paradigm for understanding why social psychology is the key to unlocking agent design.

Part 2: The Psychological Toolkit for Trust (Chapters 5–9)

With our foundations in place, we will begin to assemble our practical toolkit. This is the heart of the book, where we will translate theory into action. We will cover the step-by-step process of designing an agent's persona and onboarding flow. We will take a deep dive into the core principles of psychological influence—Reciprocity, Liking, Authority, Social Proof, Scarcity, and Consistency—and create a library of ethical design patterns for applying them in an agentic context. We will also tackle the thorny issues of transparency, explainability, and the cognitive biases that both users and designers bring to the table.

Part 3: Ethical Persuasion in Practice (Chapters 10–12)

Finally, we will put our toolkit to work. This part is about application and responsibility. We will introduce the concept of the "Guardian Agent," a wise framework for ensuring that our persuasive designs are always in service of the user's goals. We will provide a chapter dedicated to UX research methods, giving you the tools to actually measure and validate whether you are successfully building trust. We will conclude by deconstructing the persuasive strategies of real-world AI products in a series of case studies, learning from both their successes and their failures.

This roadmap is designed to be a comprehensive guide, taking you from core theory to advanced application. It is not a collection of tips and tricks. It is a new way of thinking about product design, one that places a deep understanding of human psychology at the very center of the development process.

1.5 Summary

- The central challenge in designing AI products has shifted from ensuring usability to earning a user's trust for delegation. We must move beyond "Don't Make Me Think" to answer the question, "Why should I trust you to think for me?"
- Persuasion in design is not inherently manipulative. It is the practice of ethically guiding users toward their own goals. The rise of agentic AI forces us to engage with this practice on a much deeper level.
- The **Trust Threshold** is the psychological chasm between **Usability Trust** (believing a tool works) and **Delegation Trust** (believing an agent will make good decisions on your behalf). Asking users to leap across this chasm without a bridge is a recipe for product failure.
- The cost of getting trust wrong is not just user frustration; it is product abandonment, brand damage, and wasted investment. Earning trust is a direct business imperative.
- This book provides a structured roadmap for building a bridge across.

1.6 Chapter 1 Workshop: The Trust Threshold Audit

We've covered some foundational, and frankly, quite abstract concepts in this chapter. Before we move on, let's make them concrete. The best way to understand the chasm between Usability Trust and Delegation Trust is to become an expert at spotting it in the wild—and in your own work.

This short workshop will give you a new lens to analyze digital products. The goal is to train your brain to see the Trust Threshold everywhere, so you can approach your own agentic design challenges with a clearer understanding of the psychological landscape.

Part 1: Auditing a Product You Use Every Day

How to identify the trust type.

For any feature, ask yourself one question: Who is doing the thinking?

If the user directs every action (choosing, filtering, comparing, deciding) and the product simply executes, that is Usability Trust.

If the product makes a judgment and the user accepts or rejects the outcome, that is Delegation Trust.

The shift from one to the other often coincides with a product becoming proactive: it initiates, rather than responds.

Worked example: Google Maps

Feature	Trust Type	Why
Typing a destination into the search bar	Usability Trust	The user chooses where to go. Maps executes.
Selecting a route from a list	Usability Trust	The user picks. Direct control.
"Leave now to beat traffic" suggestion	Delegation Trust	Maps inferred your schedule and made a time judgment you did not request.
"Explore" tab curating nearby restaurants	Delegation Trust	The algorithm decides what deserves your attention.
Real-time rerouting mid-drive	Delegation Trust (high)	You have handed the agent ongoing decision authority over your route.

Notice that real-time rerouting feels qualitatively different from a restaurant suggestion. The stakes, and therefore the depth of trust required, are higher. That escalation is worth paying attention to.

Your turn

Choose any feature-rich app you use regularly. Identify 5 to 7 features and sort them using the same format. The goal is not to get every answer right. It is to feel the seam between the two types of trust: the precise moment a product stops being a tool and starts acting as an agent.

Part 2: Auditing Your Own Product (15 minutes)

Worked example: A project management tool

Imagine a task-management app. Its Usability Trust features are clean and direct: creating tasks, setting due dates, attaching files. The user drives every action.

Now add an AI layer: a "Smart Prioritization" engine that automatically reorders your task list each morning based on deadlines and team activity.

Here is the trust question the team should have asked: Did we design any on-ramp for this feature? Is there a moment where users can learn how the prioritization works, correct it when it is wrong, and gradually build confidence before relying on it for real work? Or did the team assume that technical accuracy would earn trust automatically?

In most product teams, the honest answer is the second one. That gap, between a technically working feature and a psychologically trusted one, is the central design problem this book addresses.

Your turn

Apply the same lens to a product you work on. Answer these three questions:

Name three features that rely purely on Usability Trust (the user drives every action).

Identify any features that ask for Delegation Trust, however small: smart defaults, automated summaries, recommendation engines, predictive text.

For each Delegation Trust feature you found: did your team explicitly design an experience to earn that trust? Or did you assume accuracy would be enough?

There is no judgment here. The value is in the noticing.

Part 3: Imagining the Agentic Leap (20 minutes)

Worked example: The Auto-Reorder Agent

Imagine an e-commerce agent that monitors your purchase history and automatically reorders household staples, coffee, paper towels, laundry detergent, just before you run out.

What is the user's single biggest fear? Loss of financial control. Specifically: "What if it orders the wrong brand, the wrong quantity, or at the wrong moment, and charges my card before I notice?" This agent handles real money. The stakes are higher than a bad song recommendation.

What must the agent communicate on "day one" to begin earning trust? Two things: visibility ("Here is what I am planning to order, and when") and control ("You can cancel any order up to 24 hours before it ships"). Without both, even a technically perfect agent will feel threatening rather than helpful.

What is the real cost of getting trust wrong? A surprise charge. A wrong product. A cancellation and a one-star review warning others away. The financial cost is small. The trust cost is public and lasting.

Your turn

Choose one of the following features, or design your own for a product you know:

> **Project Management:** An agent that detects a looming bottleneck and proactively reassigns tasks between team members without being asked.
>
> **Mapping/Calendar:** An agent that knows your schedule and silently reroutes your morning commute, sending: "Leave 15 minutes early."
>
> **Email:** An agent that drafts and sends short, routine replies on your behalf based on your past communication style.

Answer the same three questions:

What is the user's single biggest fear? Name the exact moment they would feel their trust violated.

What must the agent communicate on "day one" to begin earning trust? Think about what you would need to hear from a new human assistant before letting them act independently.

What is the real cost of getting trust wrong? Go beyond an error message. Consider money, time, social consequences, and emotional fallout.

Keep these answers. We will return to the challenges you have identified here throughout the book as we build the psychological toolkit for addressing them.

1.7 References and Further Reading

Amershi, S., Weld, D., Vorvoreanu, M., Fourney, A., Nushi, B., Collisson, P., & Horvitz, E. (2019). Guidelines for human-AI interaction. *Proceedings of the 2019 CHI Conference on Human Factors in Computing Systems, 1-13.* Distills 20 years of HCI research into 18 actionable design principles for AI systems. It provides essential reading for operationalizing the trust-building concepts introduced in this chapter. Note: also cited in Chapter 3.

Bornet, P., Wirtz, J., Davenport, T. H., De Cremer, D., Evergreen, B., Fersht, P., Gohel, R., Khiyara, S., Sund, P., & Mullakara, N. (2025). Agentic Artificial Intelligence: Harnessing AI Agents to Reinvent Business, Work, and Life. *World Scientific Publishing*. ISBN: 9789819816224. A comprehensive, practitioner-focused guide outlines agentic AI. A panel of professionals from McKinsey, Wharton, and Babson authored the text. It directly contextualizes the business imperative for trust-building in agent design. The book argues that agents create compounding intelligence advantages if users engage with and rely on them. This idea mirrors the Trust Threshold problem this book addresses from the UX side.

Cai, C. J., Winter, S., Steiner, D., Wilcox, L., & Terry, M. (2019). Hello AI: Uncovering the onboarding needs of medical AI advisors. *Proceedings of the ACM on Human-Computer Interaction,* 3(CSCW), 1-24.

An empirical study explores how users form initial trust with AI systems. Its findings on first-interaction design provide research backing for the first date framing in the personal shopper case study in Section 1.3.

Hancock, P. A., Billings, D. R., Schaefer, K. E., Chen, J. Y. C., de Visser, E. J., & Parasuraman, R. (2011). A meta-analysis of factors affecting trust in human-robot interaction. *Human Factors, 53(5), 517-527.* A systematic review examines empirical research on trust in automated systems. The authors identify the key variables (performance, purpose, and process) that predict whether users will delegate to a machine. This directly supports the failure analysis in Section 1.3.

Huang, K. (Ed.). (2025). *Agentic AI: Theories and Practices.* Springer. eBook ISBN: 978-3-031-90026-6. An academic volume from Springer covering the development, architecture, and cross-industry implications of agentic AI systems. The author includes case studies in banking, healthcare, and cybersecurity. It provides the theoretical grounding and systems-level context that complements this book's focus on the human side of agent design.

Lee, J. D., & See, K. A. (2004). Trust in automation: Designing for appropriate reliance. *Human Factors, 46(1), 50-80.* The definitive empirical paper on trust in automated systems introduces the distinction between compliance (using a tool) and reliance (delegating to it). This forms the conceptual foundation for the Usability Trust and Delegation Trust framework in this chapter. Note: also cited in Chapter 2.

Mehrotra, S., Degachi, C., Vereschak, O., Jonker, C. M., & Tielman, M. L. (2024). A systematic review on fostering appropriate trust in human-AI interaction: Trends, opportunities and challenges. *ACM Journal on Responsible Computing, 1(4), Article 26, 1-45.* This paper covers trust in human-AI interaction broadly rather than agentic AI specifically. It stands as a comprehensive recent academic treatment of the trust-calibration problem. The authors present a 45-page systematic review of empirical methods for building appropriate trust. It provides up-to-date research grounding for the Usability Trust and Delegation Trust concepts introduced in this chapter. I included it here to complement the two agentic-AI-specific references above.

Norman, D. A. (1994). How might people interact with agents. *Communications of the ACM, 37(7), 68–71.* A classic essay by Don Norman (author of *The Design of Everyday Things*) offers a skeptical counterpoint to the agent hype of the 1990s. He argues that agents must provide a sense of control and comfort to succeed, rather than existing as strict automation. It serves as a great historical reference to show that the fear of losing control has remained a central debate for 30 years.

Parasuraman, R., & Riley, V. (1997). Humans and automation: Use, misuse, disuse, abuse. *Human Factors, 39(2), 230–253.* This paper introduces a useful taxonomy of failure. The authors distinguish between Disuse (under-reliance because the user does not trust the system) and Misuse (over-reliance because the user trusts it heavily). This supports the argument regarding the dangers of blind trust or over-delegation in the agentic era.

Shneiderman, B. (2022). *Human-Centered AI.* Oxford University Press. A comprehensive framework details how to design AI systems that improve human capability rather than replacing it. It tracks a fascinating evolution of Shneiderman's own position from the 1997 debate cited in the foundational references above.

Shneiderman, B., & Maes, P. (1997). Direct manipulation vs. interface agents. *Interactions, 4(6), 42–61.* This debate between Ben Shneiderman (advocating for user control/direct manipulation) and Pattie Maes (advocating for software agents) reflects the history of the conversation around the transition from an Interface Era to an Agentic Era described in Section 1.2.

CHAPTER 2

A Brief History of Digital Influence

To design agents that users trust, we must first understand the logic that built the web we have today. We are not starting from zero. We are building on top of two decades of behavioral science that successfully engineered the habits of billions of users.

This chapter examines the source code of modern persuasive design. We will debug the two dominant frameworks—Captology and the Hook Model—that defined the last 20 years of software development. We will identify where these models fail when applied to autonomous agents and patch them to account for the new requirement of Delegation Trust.

2.1 B.J. Fogg and Captology: The Three Ingredients of Behavior

To understand how we design for trust in agentic AI, we first have to appreciate the intellectual bedrock on which all modern persuasive design is built. That bedrock was laid down in a lab at Stanford University by a behavioral scientist named B.J. Fogg. In the late 1990s, while the rest of the world was grappling with dial-up modems and the sheer

V. Yocco, *The Persuasive Agent*, Design Thinking,
https://doi.org/10.1007/979-8-8688-3317-5_2

novelty of the World Wide Web, Fogg was asking a question that was both deeply insightful and, at the time, a little strange: Can computers persuade people?

He called his new field captology, a slightly clunky acronym for Computers as Persuasive Technologies. The core idea was revolutionary for its time. Fogg argued that computers, and digital products more broadly, were not just neutral tools for information retrieval. They were interactive systems that could be designed to influence people's attitudes and, more importantly, their behaviors.

Years later, reading through Fogg's early work, I felt a jolt of understanding. As a psychologist increasingly drawn to questions of how people interact with digital products, I knew that design choices mattered, and a language already existed to explain this in 2010. Fogg had provided a systematic, scientific framework for why they mattered. In those earlier years, the discipline that would later coalesce under the banner of "User Experience" was still finding its vocabulary: practitioners borrowed from cognitive psychology, human factors research, and information architecture, often without a shared language for what they were doing.

Fogg compressed human behavior into a logical function known as the Fogg Behavior Model. For a developer, this is the closest psychology gets to an if statement.

Figure 2-1 below summarizes how the components of the model interact with each other:

The equation is: **B = MAP**

Behavior (**B**) happens when Motivation (**M**), Ability (**A**), and a Prompt (**P**) converge at the same moment. If any variable is missing or insufficient, the function returns false, and the behavior does not occur.

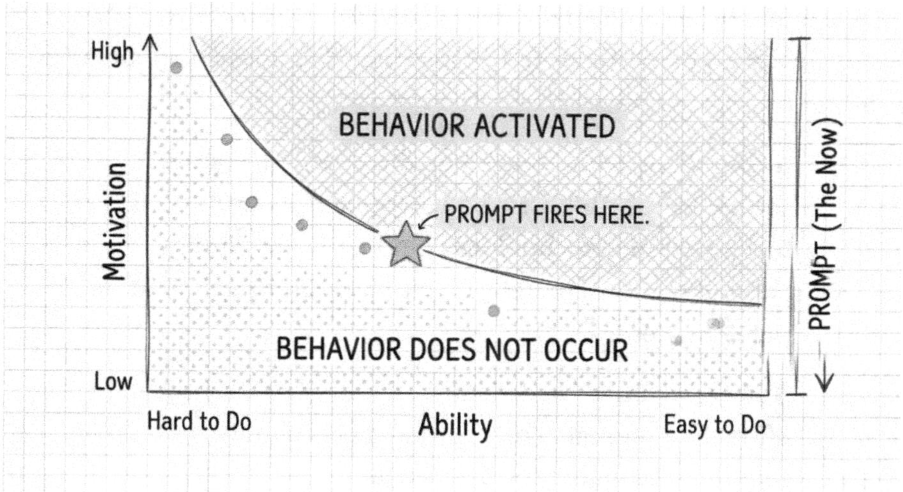

Figure 2-1. *The Fogg Behavior Model (B=MAP). Behavior occurs only when Motivation (the "why"), Ability (the "how"), and a Prompt (the "now") converge. The curved activation threshold divides the space: behaviors above the line can be triggered by a Prompt; those below cannot. The star marks an example activation point on the threshold itself, not at a theoretical maximum of both variables. Motivation and Ability trade off along the curve: a highly motivated user will accept a harder task; a nearly effortless task requires only modest motivation. The Prompt column on the right indicates that a Prompt is only effective when the user is already above the activation threshold. Image generated by author using Gemini Pro 3.1/Nano Banana 2*

Let's break that down.

- **Motivation:** This is the "why." Does the user *want* to perform the behavior?

  Fogg identifies three core motivators:

  - **Sensation:** Seeking pleasure; avoiding pain.
  - **Anticipation:** Seeking hope; avoiding fear.
  - **Belonging:** Seeking social acceptance; avoiding rejection.

A user might be motivated to buy a concert ticket because they anticipate the joy of the show, or because they fear missing out on an event their friends are attending.

- **Ability:** This is the "how." Can the user *easily* perform the behavior? Ability is about simplicity. Fogg's insight was that technology could radically simplify complex actions, thereby increasing a user's ability to perform them.

  Ability is a function of resources: time, money, physical effort, and mental effort. The less of each that a task requires, the higher the user's ability.

- **Prompt (Formerly Trigger):** This is the "now." The prompt is the cue to action. It's the click here, the notification badge, the email reminder. Without a prompt, a user who is highly motivated and has high ability might still never perform the behavior, simply because nothing tells them to do it *now*.

Fogg's model is elegant because it shows the trade-off between Motivation and Ability. If motivation is very high, people will go to incredible lengths to do something. Think of people waiting in line for hours to buy the new iPhone; motivation is so high that low ability (it's a huge pain) doesn't matter.

Ability is the lever we have the most reliable control over, which is why so much of Web 2.0 design became an obsession with simplicity and the removal of friction. The most successful products from that era ran all three B=MAP variables effectively, and whether their teams knew it or not, this model was their engine.

In this logic, our primary job is to make the desired action as simple as humanly possible. This model became the secret engine of Web 2.0. Every successful product from that era, whether they knew it or not, was a masterclass in the B=MAP formula.

- **Amazon's 1-Click Buy Button:** A masterstroke of increasing Ability. By saving your payment and shipping information, they reduced the friction of a purchase—the mental and physical effort—to virtually zero. Your Motivation to buy a book might be moderate, but because the Ability was so high, a simple Prompt (seeing the book on the homepage) was enough to trigger the behavior.
- **Facebook's Photo Tagging:** This feature tapped directly into the core motivator of Belonging. The Prompt was a notification: "You've been tagged in a photo." The Motivation was intense social curiosity and the need for self-presentation. The Ability was made simple: a single click took you right to the image. It was a perfect B=MAP storm.
- **Netflix's Autoplay:** This maximized Ability by removing the friction of decision-making. By making the next episode start automatically, they made consumption the default state and opting out the hard path.
- **Tinder's Swipe:** This maximized Ability through interface constraints. By reducing complex social evaluation to a binary gesture (left or right), they increased the speed of interaction to match the user's thought process.

For 20 years, this was the playbook. We conducted user research to understand motivations. We relentlessly simplified our interfaces to increase ability. We A/B tested our prompts to optimize conversions. In an A/B test, two versions of a design element are shown to different user segments simultaneously: version A might use a green button; version B, a blue one. The version that drives more conversions, meaning more of the desired user action (a sign-up, a purchase, a feature activation), becomes the new standard. Button colors, email subject lines, and notification timing all went through this process continuously.

We were all captologists, and B=MAP was our guiding equation. It was powerful, it was effective, and it was designed entirely for a world where the ultimate goal was to get a user to perform a direct, simple action.

2.2 The Hook Model: Building Habit-Forming Products

As the mobile era dawned and the app store exploded, the goalposts shifted. It was no longer enough to persuade a user to perform a single action. The new currency was engagement. The challenge was not just to get users to download your app, but to get them to open it again and again, ideally without any external prompting. The question became: How do you build a habit?

This is where author and behavioral designer Nir Eyal entered the scene. Building directly on Fogg's foundation, Eyal codified the process of habit formation in his hugely influential book, *Hooked*. He proposed a four-step cycle called the Hook Model, a feedback loop designed to connect a user's problem to your product with enough frequency to form an unprompted habit.

The four steps as reflected in Figure 2-2 are

1. **Trigger:** This is the "now," similar to Fogg's Prompt, but with a crucial distinction. Eyal separates triggers into two types: external and internal.

    - **External Triggers** are prompts from the environment: an email, a notification icon, a friend's recommendation. These get the user into the loop for the first time.

    - For example, you are scrolling through your phone and see a notification from a news app about a breaking story. This visual prompt gets you to open the app by grabbing your attention and tempting you to learn more about the potentially relevant topic.

- **Internal Triggers** are the ultimate goal. These are emotions, routines, or situations that cue the user to action without any external prompt. The feeling of boredom is an internal trigger for opening Instagram. The feeling of uncertainty is an internal trigger for Googling something. The product becomes mentally associated with a specific emotional state.

    - For example, you are feeling anxious about an upcoming presentation, and without consciously thinking about it, you open a meditation or relaxation app that you associate with calming down. You do this because you have successfully used the app for similar situations over a long period of time. The feeling of anxiety is the internal trigger. Your brain directly associates the app as the solution to the trigger of anxiety.

2. **Action:** This is the simplest behavior the user can take in anticipation of a reward. It builds directly on Fogg's B=MAP. The action needs to be incredibly simple (high Ability) and sufficiently motivated. For a social media app, the action is as simple as a flick of the thumb: scrolling.

- For example, TikTok or Instagram Reels:
  - **Action:** Scrolling down/flicking the thumb to view the next video.
  - **Analysis:** This action requires extremely low ability, as it is a single, easy physical movement. The motivation is high because the user is anticipating the reward of a potentially engaging, entertaining, or novel video appearing next. The design of the endless feed ensures the action is simple, immediate, and continuously reinforced by the variable reward of the content, driving the core usage loop of the product.

3. **Variable Reward:** This is the engine of the Hook. After the action, the user receives a reward, but—and this is the key—the reward is unpredictable. Drawing on the classic work of B.F. Skinner and the psychology of operant conditioning, Eyal noted that a variable schedule of reinforcement is far more compelling than a predictable one. The slot machine is the classic example. You pull the lever (Action) not knowing what you'll get. Sometimes it's nothing, sometimes it's a small win, sometimes it's the jackpot. That variability is what keeps you pulling the lever.
   - For example, on social media, the scroll (Action) delivers a Variable Reward. The next post might be boring, or it might be a fascinating article, a hilarious meme, or a picture from a close friend. You keep scrolling to see what the next "pull of the lever" will bring.

- Checking the performance of a stock in real time is a classic example of variable reward. You open your trading app or refresh the stock's page (Action) not knowing what you'll get. Most of the time the price will be stagnant or slightly down, but sometimes it's a sudden, exciting jump in value, or a significant piece of breaking news about the company. The unpredictable nature of that occasional high-value spike keeps you checking the stock price (pulling the lever).

4. **Investment:** This is the final step, where the user does a bit of work to increase the likelihood of their next pass through the hook. When you post a photo, tag a friend, or follow a new account, you are investing in the platform. This investment does two things: it loads the next trigger (you'll get notifications when people react to your post) and it makes the service better for you in the future (your feed becomes more personalized).

   - Online Banking/Personal Finance Apps (e.g., Mint, YNAB, bank apps) are an example of Investment users make in applications**:** A user takes the time to link their financial accounts, categorize transactions, and set budgets. This investment loads the next trigger (receiving alerts when they overspend) and makes the service better for them in the future (more accurate spending insights and personalized savings recommendations).

   - Loyalty Programs/Reward Apps (e.g., Starbucks, airline apps) represent user investment as well. A user takes the time to sign up for the program, save a credit card, or spend time filling out their profile with preferences. This investment loads the next trigger (receiving notifications about new deals or a free birthday item) and makes the service better for them in the future (receiving personalized offers based on their purchase history, making it faster to check out).

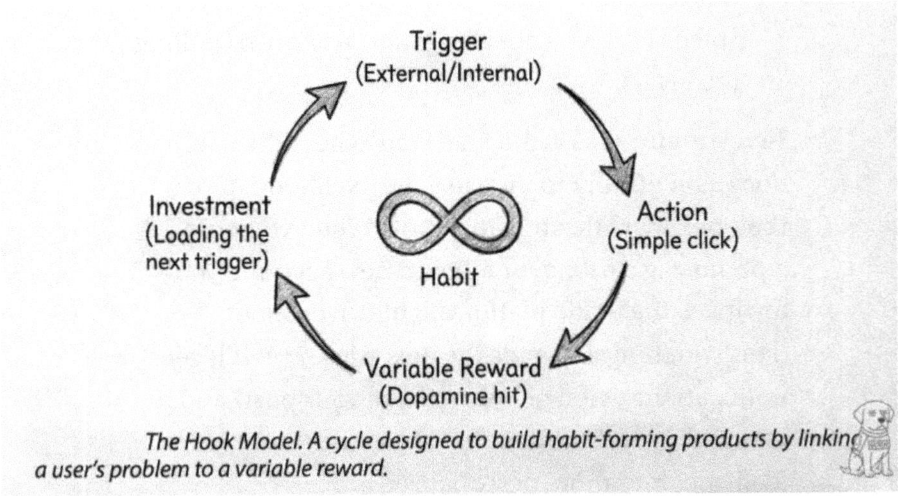

Figure 2-2. *The Hook Model expanded Fogg's model to show how digital products form habits. Image generated by author using Gemini Pro 3.1/Nano Banana 2*

The Hook Model accurately explained the meteoric rise of companies like Facebook, Twitter, and Instagram. They were masters of habit formation. They attached their products to powerful internal triggers, simplified the core action to a mindless scroll, provided an endless stream of variable rewards, and made it easy to invest and deepen the connection.

This model represented the peak of interface-based persuasion. It was the culmination of a design philosophy focused on guiding, nudging, and shaping direct user action. And like B=MAP before it, it is a powerful tool that is now rendered incomplete for the era of generative and agentic AI tools.

2.3 Why These Models Are No Longer Enough

The Fogg Behavior Model and the Hook Model are excellent frameworks for understanding how to get a person to tap a button. I've used them throughout my career, and they remain essential reading for any student of product design. But they share a critical, and now outdated, assumption: that the Behavior or Action at the core of the model is always performed *by the user.*

The entire system is designed to shepherd the user's finger to a specific point on a screen. But what happens when the most important action is no longer performed by the user at all? What happens when the action is performed *by the agent* on the user's behalf?

This is the paradigm shift that renders these models insufficient on their own. Let's revisit our agentic flight-booking assistant.

- The **Trigger** is your need to go to a meeting.
- The **Motivation** is to get a good flight without hassle.
- The simplified **Action** is not filling out a form, but simply stating your goal in natural language.
- The **Reward** is the agent returning with a perfect itinerary.

To make this concrete, consider where the classic frameworks leave us at their conclusion. At the end of a successful Hook loop, the user is engaged, motivated, and present. The B=MAP model has done its job: the Prompt has fired, the user has acted, and the reward has been delivered. But in an agentic context, the critical action has not yet happened. The flight has not been booked. The investment has not been made. The email has not been sent on your behalf. The agent has made a proposal. What happens next is determined not by interface design but by the user's willingness to delegate. Figure 2-3 maps this gap.

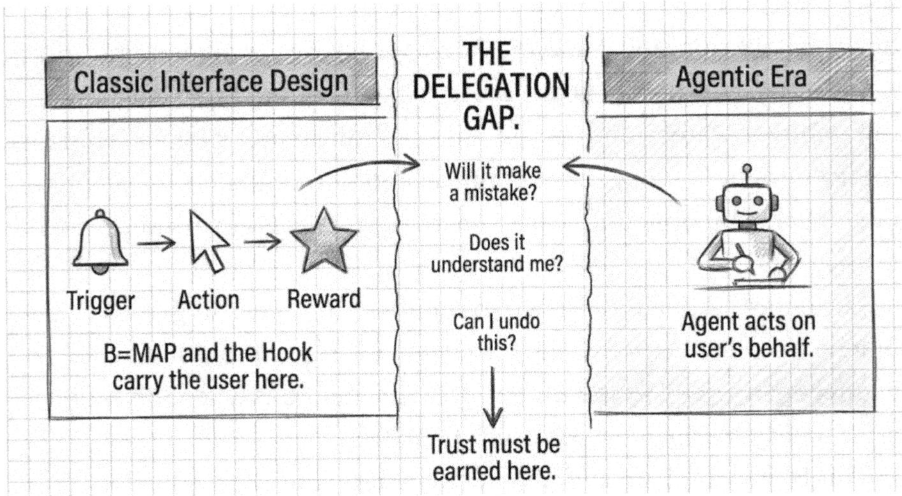

Figure 2-3. *The Delegation Gap. The classic behavioral design frameworks (B=MAP and the Hook) successfully carry a user to the edge of delegation. But the decision to accept an agent's proposal and allow it to act autonomously on their behalf happens in a zone these frameworks never mapped: the Delegation Gap. Bridging this gap is the central design challenge of the agentic era. Image generated by author using Gemini Pro 3.1/Nano Banana 2*

So far, so good. But the most critical action—the actual booking of the flight and spending of your money—is yet to come. The agent proposes, "Shall I book it?" The final, crucial behavior is you saying "yes." And your willingness to say "yes" is not a function of how easy the "yes" button is to press. Your Ability, in the classic Fogg sense, is already maxed out. Your Motivation is high. You have a Prompt.

According to the old models, this behavior should be a slam dunk. But we all know it isn't. There is a new, massive variable that has entered the equation, one that lives between the agent's suggestion and the user's consent. That variable is **Delegation Trust**.

The classic models have no language for this. They are B2C models: Business-to-Consumer. They are about the product's direct persuasion of the user. But agentic systems introduce a B2A2C model: Business-to-Agent-to-Consumer. We are now building the agent, and the agent, in turn, must persuade the consumer. Our persuasive effort is once removed, and that changes the entire dynamic.

The old models crumble because they don't account for the agent's social presence. They treat the product as an inanimate tool. But as we established in Chapter 1, users interact with agents as social actors. Therefore, the agent's ability to persuade is governed by the rules of human social psychology, not just the rules of interface usability. An agent that is technically perfect but socially clumsy, untrustworthy, or misaligned with the user's goals will fail, no matter how simple its interface is. This limitation has since been formally documented. Steyvers and Kumar (2024) identify the failure to calibrate trust as one of three core challenges in human-AI collaboration, while Weisz et al. (2024) explicitly argue that existing design guidelines, including the behavioral frameworks covered in this chapter, require new interpretations for the generative AI context.

These foundational models are not wrong. They are just the first two acts of a three-act play. They get us to the point where the user is engaged and the agent can make its proposal. But to get the user to agree to that proposal—to cross the Trust Threshold—we need to add a new layer to our thinking.

2.4 Adapting the Foundations for an Autonomous World

The good news is that we don't have to throw these models away. User-centered design has always put the person's goals ahead of the system's, and the agentic era raises the stakes because the system now acts on those goals directly. The core components: Motivation, Ability, Prompts, Rewards, Investment, are still fundamental to human behavior. We just need to upgrade our understanding of them for the agentic era. We need to see them through the lens of Delegation Trust. Table 2-1 shows the changes needed to the components to address the shortcomings of these models from the past.

Table 2-1. *Changes needed to address our shortcomings from the historical non-AI interfaces*

Component	The Interface Model (Old)	The Agent Model (New)
Motivation	Motivate Action: Convince the user to perform the task.	Motivate Delegation: Convince the user to let go of the task. We must address the fear of losing control.
Ability	Simplicity of Action: Make the button easy to click. Reduce physical friction.	Simplicity of Trust: Make the decision to trust easy to make. Provide transparency, reversibility (undo), and clear logic.
Reward	Content/Social: Likes, information, or entertainment.	Competence: The wow moment is competence-by-proxy. The agent did it as well as (or better than) the user could have.
Investment	Data Entry: User adds content to customize the feed.	Teaching: User corrects the agent. This is not debugging; it is training. It builds a proprietary bond between user and agent.

As shown in the table there are meaningful changes required for each element from our old models:

Motivation Becomes Motivating Delegation

- **Old Model:** How do we motivate the user to *do the task*?
- **New Model:** The agent does the task. Our new job is to motivate the user to *let go of the task*. This means addressing their fears and anxieties. The primary motivation we must tap into is the desire to reduce cognitive load, save time, and achieve a goal with less effort. But we must also overcome the counter-motivating forces of fear: fear of losing control; fear of the agent making a mistake; fear of the agent not understanding our intent; and fear of surrendering sensitive personal data, including financial records, location history, and private communications, to third-party AI systems whose data practices may be unclear. This last fear deserves particular attention in agentic design. An agent that needs access to your calendar, your inbox, or your bank account is asking for a level of access that most users have never previously granted to software. Acknowledging this directly, and offering meaningful transparency about data use and retention is a trust-building act.

Ability Becomes Simplicity of Trust

- **Old Model:** How do we make the action as simple as possible to *perform*? (e.g., 1-Click Buy)
- **New Model:** How do we make the decision to trust as simple as possible to *make*? This involves things like

- **Clarity:** Providing just enough information about how the agent reached its conclusion.
- **Reversibility:** Offering a clear undo or cancel option to lower the stakes of a bad decision.
- **Scaffolding:** Starting with small, low-risk delegations and gradually asking for more trust over time as the agent proves its competence.

The mechanism here extends beyond transparency alone. Metzger et al. (2024) found in a controlled study that limitation disclaimers had no significant effect on user trust, while an authoritative communicative style did increase it, suggesting that tone and persona are more powerful trust levers than information disclosure.

The Reward Becomes the Competence Loop

- **Old Model:** The reward is often social validation (a like), new information, or a solved problem.
- **New Model:** The most powerful reward an agent can deliver is the feeling of *competence-by-proxy*. The "wow" moment is not just that the task is done, but that it was done *well,* just as you would have done it, or even better. Every time an agent successfully completes a delegated task, it's not just a reward; it's a deposit in the trust bank. This creates a powerful feedback loop: successful delegation is the reward that encourages more delegation.

Investment Becomes Teaching and Correction

- **Old Model:** Investment is about adding content or data to personalize an experience.
- **New Model:** The most valuable investment a user can make in an agent is teaching it. When a user corrects an agent's mistake or provides clarifying feedback ("No, not that kind of restaurant, I meant one that's better for a business lunch"), they are not just fixing an error. They are actively training their personal agent. A well-designed agent makes this process feel rewarding, framing it as a collaborative act of teaching, not a frustrating act of debugging. This investment builds a powerful bond and makes the agent uniquely valuable to that specific user.

Notice that the updated components described above do not form a closed loop, the way the original Hook Model does. Instead, they describe an expanding spiral. Each successful delegation is a deposit in the trust bank. The agent proves its competence, the user extends a little more autonomy, the agent performs at that new level, and the trust bank grows again. This is the Agentic Trust Spiral: a virtuous cycle in which a well-designed agent and an increasingly confident user grow toward each other over time.

The crucial difference from the Hook is directionality. The Hook loops back to zero: a user's relationship with a social media app can remain flat for years, cycling through the same action and reward without deepening. But a user's relationship with a well-designed agent should grow over time, moving from small, low-stakes delegations ("Summarize this article") toward increasingly significant ones ("Book my travel," "Draft my proposal") as trust accumulates. Figure 2-4 maps this progression.

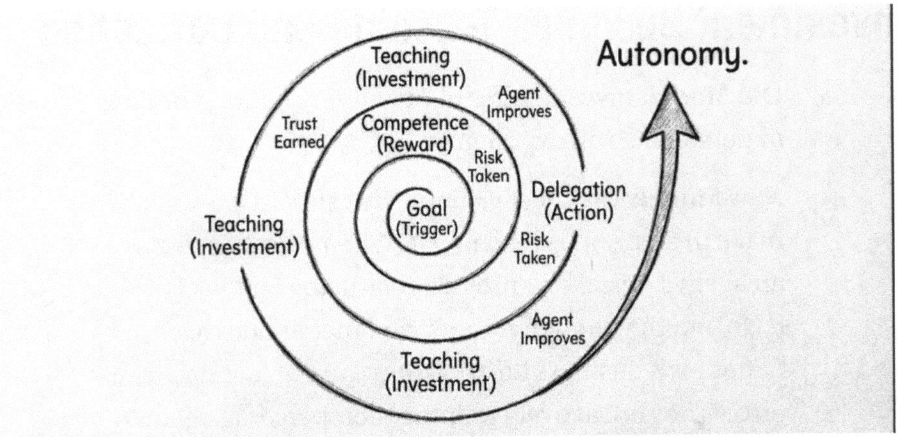

Figure 2-4. *The agentic trust spiral. Unlike a closed habit loop, agentic interaction is an expanding spiral. Each successful delegation builds trust and improves the agent, leading to greater autonomy*

2.5 Summary

- The intentional design of digital influence has a history. Understanding this history gives us the context we need to tackle the new challenges of AI.
- B.J. Fogg's Behavior Model (B=MAP) gave us the foundational equation for interface-based persuasion: Behavior happens when Motivation, Ability, and a Prompt converge. It taught us that simplifying an action is one of the most powerful persuasive tools.
- Nir Eyal's Hook Model (Trigger, Action, Variable Reward, Investment) evolved this thinking for the mobile era, providing a framework for building user habits by connecting a product to an internal trigger.

- These powerful models are no longer sufficient because they were built for a world where the user always performs the critical action. They have no language for Delegation Trust, the essential ingredient for agentic systems.
- We must adapt these foundational concepts. Our goal shifts from motivating action to motivating delegation, from making the task simple to making the act of trusting simple, and from providing rewards to building a loop of proven competence.

We now have our historical context. We understand the intellectual shoulders we are standing on, and we see the limits of the old worldviews. With this foundation in place, we are ready to introduce the first of our new tools for designing agentic interactions: a robust framework for improving communication between humans and their agents.

2.6 Chapter 2 Workshop: Deconstructing Influence

A Guided Analysis

The frameworks in this chapter are most powerful when you can see them operating in products you already use every day. This workshop guides you through three steps: mapping a product's persuasive engine using the Hook Model; locating the exact point where that engine runs out of road; and sketching what trust-centered design might look like instead.

We will work through a complete example together first. Then you will apply the same process to a product of your choice.

Worked Example: TikTok

TikTok is one of the most studied examples of habit-forming design. Here is its Hook mapped out precisely.

Step 1: Map the Hook

Trigger. The internal trigger is boredom: the low-grade discomfort of having nothing engaging to occupy attention. External triggers include the notification badge on the app icon and push notifications such as "New videos just for you."

Action. A single upward flick of the thumb loads the next video. TikTok has maximized Ability to its logical limit: one gesture, zero decisions, zero cognitive load required.

Variable Reward. The next video might be unremarkable, or it might be hilarious, emotionally resonant, or professionally useful. The uncertainty is the point. This is Skinner's variable reinforcement schedule running in pure digital form.

Investment. Watching a video to the end, liking it, following the creator, or sharing it all train the algorithm. Your feed becomes more calibrated to your preferences, making future sessions more rewarding and automatically loading the next internal trigger.

Step 2: Find the Delegation Gap

TikTok's core feed stays largely within B=MAP territory: scrolling through videos involves no act of delegation. But TikTok has expanded into TikTok Shop, where a buy button embedded in a creator's video allows you to purchase a product without leaving the app.

Here, the Delegation Gap appears clearly. The Hook brings you to the video and the button. The B=MAP model explains why tapping the button is nearly effortless. But completing the purchase requires something fundamentally different: you need to trust that the product is as shown, the seller is legitimate, and TikTok's payment system will work correctly. That calculation is not about friction. It is about Delegation Trust.

The user's trust calculus, spoken aloud, sounds like this: "I must trust that this seller is real, that the product will arrive as shown in the video, and that my payment information is handled securely by a platform I primarily associate with entertainment."

Step 3: Upgrade the Design

Using the adapted frameworks from Section 2.4, here are three targeted design changes:

Motivating Delegation. Show a "Verified Seller" badge alongside the buy button, paired with a count of recent purchases ("1,200 sold this week"). This directly reduces the fear of the unknown seller, the primary trust barrier in this context.

Simplicity of Trust. Add a post-purchase cancellation window: a "Cancel within 5 minutes" option displayed immediately after the purchase confirmation screen. This lowers the stakes of the delegation. The user is making a revisable commitment, not an irreversible one.

Investment as Teaching. After delivery, prompt the user with "How well did this match what you saw in the video?" rather than a simple star rating. This frames feedback as refining the recommendation system rather than rating a transaction.

2.7 Your Turn: A Guided Analysis

Now apply the same three steps to an AI-powered feature of your own choosing. Good candidates include:

Spotify's Discover Weekly or Made for You playlists

Netflix's Top Picks for You recommendations

An AI coding assistant such as GitHub Copilot or Cursor

A smart assistant offering a proactive suggestion (Siri, Google Assistant)

Step 1: Map the Hook

For the feature you chose, identify and write down

- **The Internal Trigger**: What emotional state or recurring situation prompts the user to engage?
- **The External Trigger**: What notifications, badges, or cues attract the user's attention?

- **The Action**: What is the simplest behavior the user performs in anticipation of a reward? How has the company made this action nearly frictionless?
- **The Variable Reward**: What is the unpredictable payoff? Is it social, informational, or self-oriented (a sense of progress or mastery)?
- **The Investment**: What does the user do that makes the product more valuable to them in the future and loads the next trigger?

Step 2: Find the Delegation Gap

Now identify the moment where the Hook logic stops. Ask yourself

- What is the simple initial action the Hook describes?
- What is the deeper, more meaningful act of delegation the product is actually asking for? (Hint: it is usually the moment the user must accept the AI's judgment and relinquish direct control.)
- Write out the user's trust calculus in plain language. Complete this sentence at least three times: "To use this feature fully, I must trust that..."

Step 3: Upgrade the Design

Propose one specific design change for each of the three updated principles:

Motivating Delegation. Name one specific fear from the user's trust calculus and describe a design element that addresses it directly.

Simplicity of Trust. How could you lower the stakes of the delegation? Consider reversibility (can the user undo the agent's action?), clarity (does the user understand why the agent made its recommendation?), or scaffolding (can the user start with a low-stakes delegation and build up gradually?).

Investment As Teaching. How could you reframe one feedback or correction mechanism so that users feel they are training their personal agent rather than debugging a broken tool?

Keep These Notes. The vocabulary you are developing here: trust calculus, delegation gap, simplicity of trust, is the vocabulary we will use for the rest of the book.

2.8 References and Further Reading

Eyal, N. (2014). *Hooked: How to build habit-forming products.* Portfolio. This book introduces the "Hook Model," a four-step cyclical process (Trigger, Action, Variable Reward, Investment) designed to explain how products create unprompted user engagement. Eyal builds on earlier behavioral science to demonstrate how successful companies link their products to internal emotional triggers, effectively manufacturing habits.

Fogg, B. J. (2003). *Persuasive technology: Using computers to change what we think and do.* Morgan Kaufmann. In this seminal text, Fogg defines the field of "captology" (Computers as Persuasive Technologies), arguing that digital tools are not neutral but are interactive systems capable of influencing user attitudes and behaviors. This work laid the intellectual bedrock for modern persuasive design, moving the field beyond simple usability and into the realm of psychological influence.

Fogg, B. J. (2009). A behavior model for persuasive design. *Proceedings of the 4th International Conference on Persuasive Technology*, 40. This paper formally introduces the Fogg Behavior Model (B=MAP), which posits that behavior only occurs when Motivation, Ability, and a Prompt converge at the same moment. It provides the "source code" for interface-based persuasion, teaching designers that increasing "Ability" (simplicity) is often the most effective way to drive user action.

Lee, J. D., & See, K. A. (2004). Trust in automation: Designing for appropriate reliance. *Human Factors*, 46(1), 50–80. This paper is the gold standard for understanding Trust in automated systems. It introduces the distinction between *compliance* (using a tool) and *reliance* (delegating to an agent), which supports my argument about the shift from Usability Trust to Delegation Trust.

Matz, S. C., Teeny, J. D., Vaid, S. S., Peters, H., Harari, G. M., & Cerf, M. (2024). The potential of generative AI for personalized persuasion at scale. *Scientific Reports*, 14, 4692. Across four studies (roughly 1,788 participants), personalized messages written by ChatGPT influenced people significantly more than generic ones, across product marketing, political appeals, and several psychological profiles, even from a single short prompt naming the target trait. This study extends the Fogg and Eyal lineage into the present and gives Sections 2.3 and 2.4 concrete evidence that the persuasion map has changed: what once took a marketing team can now run automatically at scale.

Metzger, L., Miller, L., Baumann, M., & Kraus, J. (2024). Empowering calibrated (dis-)trust in conversational agents: A user study on the persuasive power of limitation disclaimers vs. authoritative style. *Proceedings of the 2024 CHI Conference on Human Factors in Computing Systems* (CHI '24), Article 481. A controlled study comparing two ways to shape trust in a conversational agent: stating the agent's limitations versus adopting an authoritative style. The disclaimers showed no significant effect, while the authoritative style raised trust.

Reeves, B., & Nass, C. (1996). *The media equation: How people treat computers, television, and new media like real people and places.* Cambridge University Press. This is the definitive text for the CASA (Computers Are Social Actors) paradigm. It provides empirical evidence that humans unconsciously apply social rules (politeness, reciprocity) to machines, which is the psychological basis for why an agent's personality matters.

Steyvers, M., & Kumar, A. (2024). Three challenges for AI-assisted decision-making. *Perspectives on Psychological Science,* 19(5), 722–734. This paper names three obstacles that stand between human-AI teams and their full potential, with the failure to calibrate trust chief among them. It gives Section 2.3 peer-reviewed backing for the argument that the classic behavioral models stop short: they carry a user to the edge of delegation but say nothing about whether the user should trust the agent waiting on the other side.

Wang, L., Ma, C., Feng, X., Zhang, Z., Yang, H., Zhang, J., ... Wen, J.-R. (2024). A survey on large language model based autonomous agents. *Frontiers of Computer Science,* 18(6), 186345. The first peer-reviewed survey published in this field, presenting a unified framework for the construction, application, and evaluation of autonomous LLM agents.

Weisz, J. D., He, J., Muller, M., Hoefer, G., Miles, R., & Geyer, W. (2024). Design principles for generative AI applications. *Proceedings of the 2024 CHI Conference on Human Factors in Computing Systems* (CHI '24), Article 378, 1–22. Weisz and colleagues present six design principles for generative AI applications, developed through literature review, practitioner feedback, and validation against real products. They argue that older guidelines, including the behavioral frameworks covered in this chapter, need fresh interpretation once a system generates content and acts on a user's behalf. This supports the point that Fogg and Eyal remain useful yet incomplete for agentic design.

CHAPTER 3

Debugging the Conversation: A Communication Framework for Agent Design

You have just shipped a technically perfect AI feature. The response engine is accurate. The workflow logic is sound. Internal testing was flawless. And then the real data arrives.

Users are engaging once, frowning at their screens, and leaving. The drop-off is not at signup. It is happening right in the middle of conversations, at the exact moment the agent does something technically correct but humanly wrong.

The post-mortem begins. Clean logs. No bugs. And then someone in the room says those four words that every design team dreads: "It just doesn't feel intuitive."

The room goes quiet. There is nothing to refactor. Nothing to hot-fix. Just a ghost in the machine that nobody can name.

V. Yocco, *The Persuasive Agent*, Design Thinking,
https://doi.org/10.1007/979-8-8688-3317-5_3

We traced the history of this problem in the previous chapter. We saw how frameworks built for getting users to tap buttons must now stretch to cover something far more complex: earning the kind of trust that gets a user to say, "Yes, go ahead and do it for me." But knowing the history is not enough. You need a set of tools, not just a map.

When an API fails, you have a stack trace. You can look at the logs, identify the timeout or the bad request, and apply a fix. When a human-agent interaction fails, the logs are often messy, emotional, and subjective. Users say things like, "It feels weird," or "I don't trust it."

Our job in this chapter is to change that. We need to turn these subjective feelings into a debuggable system. We need a schematic for conversation that allows us to spot errors with the same precision we apply to our code. This chapter introduces that schematic: The Transactional Model of Communication.

3.1 Beyond Intuitive: A Formal Model for Interaction

I want you to picture a scene that will be familiar to anyone who has worked in tech for more than a week. We're in a design critique meeting. A junior designer is presenting a new workflow for a complex feature. They finish the walkthrough, and the room is quiet for a moment. Then, a senior stakeholder clears their throat and offers the most famously unhelpful piece of feedback in the corporate lexicon:

"I don't know... it just doesn't feel very *intuitive*."

The word hangs in the air, a fog of subjectivity. What does it even mean? For the stakeholder, it's a genuine expression of their experience. For the engineering and product teams, it is a dead end. *Intuitive* is a subjective feeling, not a technical requirement. There is nothing to refactor. The conversation devolves into a guessing game. Do they mean the button is in the wrong place? Is the copy confusing? Should we use a different color? We lack a shared, precise vocabulary to diagnose the problem, so we poke at the surface, hoping to stumble upon a solution.

For years, *intuitive* has been the highest praise we could bestow upon an interface. We've used it as a shorthand for "easy to use without instruction." While well-intentioned, this reliance on vague, subjective terms is a luxury we can no longer afford. When designing simple, direct-manipulation interfaces, we could often get away with it. But when we are building conversational agents that engage in complex, multi-turn interactions, "it doesn't feel intuitive" is a catastrophic failure of analysis.

A failed conversation with an AI agent is not a simple usability issue. It is a communication breakdown. And to fix a communication breakdown, you need a **model of communication**. You need a way to deconstruct the interaction into its fundamental components so you can pinpoint the exact point of failure. You need to move beyond guesswork and into diagnostics.

Fortunately, we don't have to invent this. The field of communication studies has been analyzing the mechanics of human interaction for nearly a century. Scholars have developed powerful models to explain how meaning is created, shared, and tragically, lost between two parties. For our purposes, the most useful of these is the **Transactional Model of Communication**.

This model provides the shared, precise vocabulary we've been missing. It gives us a schematic, a wiring diagram of a conversation, allowing us to see how all the pieces connect. It will allow us to replace "it feels confusing" with "I believe we have a semantic noise problem in how the agent is encoding its response." That may sound overly academic, but that level of precision is exactly what's required to solve these new, complex interaction problems. In this chapter, we will unpack this model and turn it into the primary analytical tool in your agent-design toolkit.

3.2 Deconstructing Conversation: Senders, Receivers, Messages, and Channels

The Transactional Model of Communication views conversation not as a simple, linear act of one person sending a message to another but as a simultaneous, dynamic process where both participants are continuously sending, receiving, and interpreting signals. Think of it less like a game of catch and more like a dance. Both partners are moving at the same time, reacting to each other's subtle cues in a constant feedback loop. Figure 3-1 shows how this interplay works in theory.

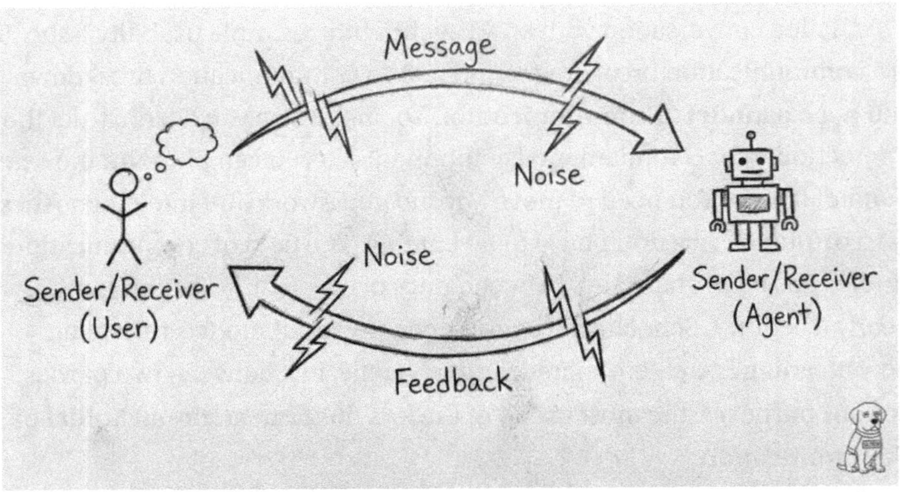

Figure 3-1. *The Transactional Model of Communication. The model illustrates conversation as a dynamic, simultaneous process. Both participants (User and Agent) act as both* ***Sender*** *and* ***Receiver****, continuously* ***Encoding*** *(creating) and* ***Decoding*** *(interpreting)* ***Messages*** *through a* ***Channel****. Image generated by author using Gemini Pro 3.1/Nano Banana 2*

Let's break down the core components of this dance.

Senders and Receivers

In any communication, there is a sender (the one initiating a message) and a receiver (the one interpreting it). In the transactional model, these roles are fluid. The moment you respond to someone, you switch from receiver to sender. In a human-agent interaction

- When the user types a prompt, the **user is the sender** and the **agent is the receiver**.
- When the agent displays its response, the agent is the sender and the user is the receiver.

  This constant, fluid exchange is the basic rhythm of conversation.

Message

The message is the *what* of the communication. It's the collection of symbols—words, images, sounds—that represent the sender's thought. The message is more than just the literal text. It has layers:

- **Content:** The literal meaning of the words.
- **Relational information:** The tone and style that suggest the nature of the relationship between the sender and receiver. "Submit your query" and "What can I help you with today?" may have the same functional content, but their relational messages are worlds apart.

  The agent's message is not just the text it generates. It's the UI that contains the text, the micro-animations, the loading indicators; everything the user perceives is part of the message.

Encoding and Decoding

This is the most critical, and often most problematic, part of the process.

- **Encoding** is the sender's process of converting a thought into a message. You have an idea in your head, and you encode it into spoken words, typed text, or body language.
- **Decoding** is the receiver's process of interpreting that message and converting it back into a thought.

Every communication failure is, at its heart, an encoding/decoding failure. The thought in the sender's head did not perfectly match the thought created in the receiver's head. When you write a prompt for an AI, you are encoding your complex goal into a string of text. The AI then decodes that text to interpret your intent. Its entire response is an attempt to encode a solution back to you. We spend the rest of this book focused on making that encoding/decoding process more reliable.

Channel

The channel is the medium through which the message travels. Is it a text-based chat window? A voice interface? A graphical display? A combination of all three? The choice of channel dramatically affects the message. A sarcastic comment that is obvious when spoken (channel: voice, with tonal cues) can be easily misinterpreted when read (channel: text, without tonal cues). A key part of agent design is choosing the right channel for the right message. A complex data table should probably be displayed visually, not read aloud by a voice agent.

Feedback

Feedback is the receiver's response to the sender's message. It is the part of the dance where you find out if your partner is following your lead. Feedback can be verbal (I don't understand) or non-verbal (a confused look). In a GUI, a click is a form of feedback. In an agentic system, a user correcting the agent (No, I meant Paris, Texas, not Paris, France) is a crucial feedback signal. Well-designed agents make it easy for users to provide this corrective feedback.

Field of Experience

This is perhaps the most important concept for understanding AI failures. Each participant in a conversation brings their own unique field of experience—their culture, their past experiences, their personal knowledge, their emotional state. Communication is most effective when those fields of experience overlap. When you and a close friend use an inside joke, you are operating within a massive shared field of experience.

An AI agent has no real-world experience. Its experience is its training data. A user, on the other hand, is a universe of unique context. They might be stressed, in a hurry, using a colloquialism the AI has never seen, or operating with a set of cultural assumptions that are absent from the training data. As Figure 3-2 shows us, the tiny sliver where the user's field of experience overlaps with the AI's training data is the only place where successful communication can happen. Our job is to make that overlap as large as possible.

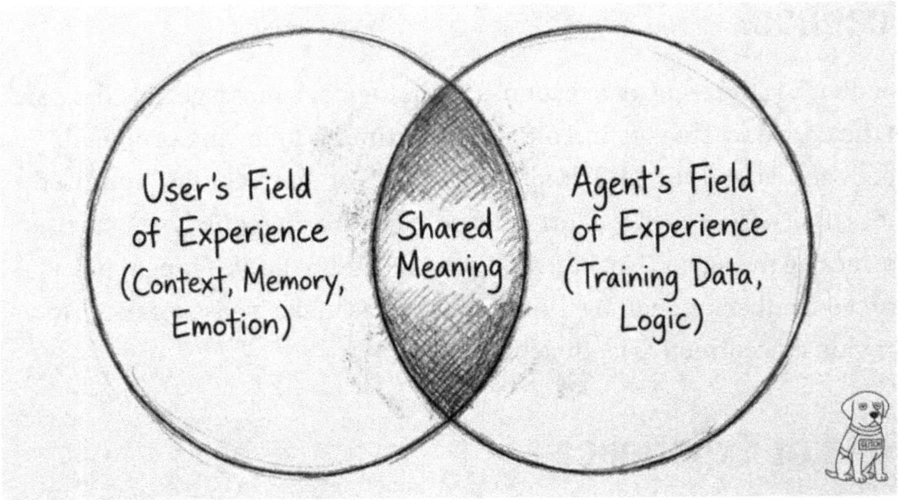

Figure 3-2. *The Field of Experience Overlap. Successful communication (the area of overlap) only occurs when the user's personal context and expectations align with the AI's limited training data and capabilities. Our job is to design interactions that expand this shared context and minimize the gap. Image generated by author using Gemini Pro 3.1/Nano Banana 2*

3.3 Finding the Noise: Technical, Semantic, and Psychological Interference

So, if communication is a dance, what makes us step on each other's toes? In the transactional model, any interference that disrupts the fidelity of the message is called **noise**. Noise is the reason the thought in the sender's head doesn't perfectly match the thought in the receiver's head.

This is where the model becomes a powerful diagnostic tool. By identifying the *type* of noise that is occurring, we can move from "it's confusing" to a precise diagnosis. There are three types of noise. Figure 3-3 shows how noise can serve to deteriorate the intended message on its way from sender to receiver. Our job is to create experiences that reduce noise, accounting for as many potential blockers of the message as possible.

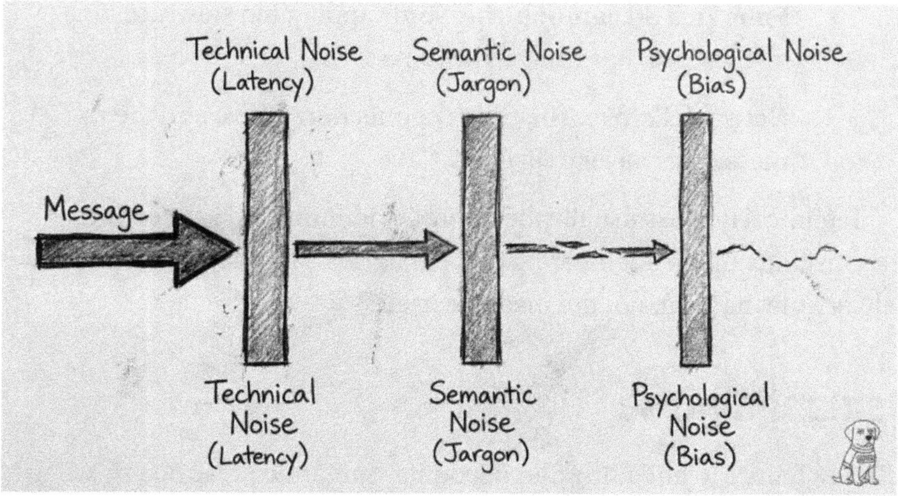

***Figure 3-3.** The Three Types of Noise. Technical noise blocks the signal, semantic noise corrupts the meaning, and psychological noise distorts the interpretation. Image generated by author using Gemini Pro 3.1/Nano Banana 2*

Technical Noise

This is the simplest form of interference. Technical noise is any physical disruption in the channel. A bad cell phone connection, a frozen video call, or a garbled microphone are all examples of technical noise. The sender and receiver might be perfectly aligned, but the message itself gets corrupted in transit.

In human-agent interaction, technical noise includes

- **API Latency:** The agent takes too long to respond, making the conversation feel stilted and unnatural.
- **A Buggy UI:** The chat window glitches, text overlaps, or the scroll position jumps erratically.

- **Poor Text Streaming:** The words appear too slowly or in awkward, hard-to-read chunks.
- **Network Errors:** The user's connection drops, and the message is never delivered.

Technical noise is usually the easiest to identify and is often the responsibility of the engineering team to fix. It's a problem with the delivery mechanism, not the message itself.

Semantic Noise

This is a failure in encoding and decoding. Semantic noise occurs when the sender and receiver have different interpretations of the symbols used in the message. Jargon, ambiguous language, and cultural idioms are common sources of semantic noise. If a doctor tells a patient they have "idiopathic rhinitis," the patient might be terrified. If they say "we don't know what's causing your runny nose," the message is clear. Same content, different encoding.

In human-agent interaction, semantic noise is rampant:

- **The User's Vague Prompt:** A user asks, "Find me a good Italian restaurant." The word "good" is semantically noisy. Does it mean cheap? Authentic? Highly rated? Close by?
- **The Agent's Confusing Language:** The AI responds with an overly technical or verbose sentence that obscures the core answer.
- **Misaligned Vocabulary:** A user asks a financial AI to short a stock, but the AI was trained on more conservative investment language and doesn't understand the slang.

- **Ambiguity:** The agent's response could be interpreted in multiple ways. Your package will arrive soon. How soon is soon? An hour? Two days?

Semantic noise is a design and content problem. Solving it involves training the AI to be clearer in its responses and better at asking clarifying questions when it receives an ambiguous prompt.

Psychological Noise

This is the most complex, subtle, and powerful form of interference. Psychological noise refers to the internal, cognitive, and emotional states of the participants that color their perception of the message. It's the internal monologue that is running while you're trying to listen.

If you've had a bad day, you might interpret a neutral email from your boss as hostile. If you are a deep expert in a topic, you might find a simplified explanation to be condescending. These are examples of psychological noise.

In human-agent interaction, psychological noise is the invisible force behind most trust failures:

- **Pre-existing Bias:** A user who is already skeptical of AI will scrutinize every response for errors, while a user with strong automation bias will accept a flawed answer without question.
- **Emotional State:** A user who is stressed or in a hurry will have very little patience for a chatty, verbose agent. A relaxed, exploratory user might appreciate that same personality.
- **Misaligned Persona:** If an agent designed to be a fun, creative muse is used for a serious legal research task, the psychological mismatch between the user's goal

and the agent's persona will create constant friction. The user's psychological state (I need to be serious and focused) is at odds with the agent's.

- **Lack of Context:** The agent doesn't know that the user just had a bad sales call, and its cheerful, "How can I make your day great?" message lands with a thud.

Psychological noise is the hardest to solve, but it's also where the biggest gains in user trust can be found. It requires us to think beyond the conversation itself and consider the user's entire context, their goals, and their emotional state. It requires us to design the agent's persona and communication style with extreme care. Reducing psychological noise is user-centered design in practice: you study the real person on the other side of the screen and adapt the agent to their state instead of asking them to adapt to yours.

Table 3-1 consolidates these three noise types into a quick-reference diagnostic guide. Think of it as the bug report template you can use in any design critique or sprint retrospective.

Table 3-1. *Debugging communication noise*

Noise Type	User Symptom (The Bug Report)	Root Cause (The Error Log)	Debugging Strategy (The Patch)
Technical	It's slow, It's broken, The button didn't work.	Channel Failure: Latency, network timeouts, or rendering glitches are breaking the flow.	Optimize Performance: specific load states, optimistic UI updates, or retry logic.
Semantic	It didn't understand me, It gave me the wrong answer, Why did it say that?	Encoding/Decoding Error: The agent's parser failed to map the user's vague words to specific variables.	Clarification Loops: Implement "disambiguation" flows where the agent asks for specifics before acting.
Psychological	It feels creepy, It's annoying, I don't trust it, It was rude.	State Mismatch: The agent's tone or actions clashed with the user's emotional state or expectations.	Persona Adjustment: Adjust the agent's "temperature" or formality based on sentiment analysis or context.

3.4 Using the Model to Diagnose a Failed AI Interaction

Let's put our new tool to work. Imagine a user, Sarah, is interacting with a new AI-powered project management assistant. Here's the transcript:

> **Sarah:** Hey, can you push the deadline for the Q4 launch presentation back a bit? Things are running behind.

Agent: Affirmative. The deadline for the Q4 Launch Presentation task has been moved. The new due date is October 24th, 2025.

Sarah: Whoa, no! That's only one day! I need more time. Push it back a week.

Agent: I have updated the task. The new due date is October 31st, 2025.

Sarah: (Now frustrated, opens the calendar manually and changes the date herself).

This interaction feels clunky. It isn't intuitive. But now, we have the tools to do better. Let's diagnose this failure using our model of communication

1. **Identify the Senders, Receivers, and Messages.**
    - **Turn 1:** Sarah is the sender. Her message is a request to delay a deadline.
    - **Turn 2:** The agent is the sender. Its message is a confirmation of an action.
2. **Pinpoint the Noise.** Let's look for interference in Sarah's first message: *push the deadline back a bit? Things are running behind.*
    - **Semantic Noise:** The phrase "a bit" is incredibly noisy. It is a classic example of ambiguous language. Sarah encoded her thought (I probably need about a week) into a vague phrase. The agent decoded "a bit" based on its training data, which might interpret it as the smallest possible unit: one day. This is the primary point of failure.

- **Psychological Noise:** The agent's persona is also a problem. Its response, *Affirmative*, is cold, robotic, and formal. Sarah's initial message was informal and slightly stressed (Things are running behind). The agent's formal tone creates a psychological mismatch. It doesn't acknowledge her context or emotional state. It feels like she's talking to a machine that doesn't understand the human reality of a project slipping. This erodes trust and makes the semantic error even more jarring.

- **Technical Noise:** There appears to be no technical noise. The responses were fast and the UI worked.

3. **Prescribe a Solution.**

   Our diagnosis points to clear solutions. We have a semantic problem and a psychological one.

   - **To Solve the Semantic Noise:** The agent should be trained to recognize ambiguous temporal phrases like "a bit," "later," or "soon." When it detects one, it should not take action. Instead, it should ask a clarifying question to resolve the ambiguity.

   - **To Solve the Psychological Noise:** The agent's persona needs to be adjusted to feel more collaborative and less robotic. It should acknowledge the user's context.

Consider this new version of the interaction, designed directly from our diagnosis:

> **Sarah:** "Hey, can you push the deadline for the Q4 launch presentation back a bit? Things are running behind."

Agent (New Version): It sounds like things are tight on the Q4 launch. I can definitely move the deadline. To make sure I get it right, how much more time do you need? A few days, or a week?

Sarah: Let's do a week to be safe.

Agent (New Version): Okay, I've moved the deadline to October 31st. Hope that gives you the breathing room you need!

Notice the difference. The new agent acknowledges Sarah's psychological state (It sounds like things are tight). It resolves the semantic noise by asking a clarifying question instead of making a bad assumption. The final confirmation is encouraging and aligned with a more supportive, collaborative persona. This interaction builds trust. The original one destroyed it. We were only able to arrive at this better design because we had a formal model to diagnose the specific points of failure.

3.5 Summary

- To design effective AI agents, we must move beyond vague, subjective terms like "intuitive" and adopt a more rigorous, diagnostic approach to interaction design.
- The **Transactional Model of Communication** provides a powerful framework for this. It deconstructs conversation into its core components: Senders, Receivers, Messages, Channels, Feedback, and Fields of Experience.
- Encoding/Decoding is the serialization of intent into text. This is where most errors occur.

- Fields of Experience define the shared context. AI has very little context, so we must design mechanisms to capture it.
- A communication breakdown is caused by **noise**, which prevents the sender's intended meaning from being successfully recreated in the receiver's mind.
- We can diagnose interaction failures by identifying three types of noise: **Technical Noise** (a problem with the channel), **Semantic Noise** (a problem with ambiguous symbols), and **Psychological Noise** (a problem with internal biases and emotional states).
- Technical Noise is a channel failure (latency, bugs).
- Semantic Noise is a symbol failure (ambiguity, jargon).
- Psychological Noise is a state failure (bias, emotion, wrong persona).
- By using this model as a diagnostic tool, we can systematically identify the root causes of failed human-agent interactions and design solutions that are more robust, collaborative, and trustworthy.

We now have our first major tool. We have a lens through which we can analyze any conversation. But this model raises a crucial question. Why does it even work? Why do we analyze a conversation with a piece of software as if it were a person? The answer lies in a fascinating and fundamental aspect of human psychology, which we will explore in the next chapter.

3.6 Chapter 3 Workshop: The Communication Autopsy

- In this chapter, we gave you a new vocabulary: Senders, Receivers, Channels, Fields of Experience, and three distinct types of noise. The best way to internalize a vocabulary is to use it. In this guided exercise, you will perform a communication autopsy on a failed AI interaction, moving from a vague cause of death to a precise understanding of what went wrong. We will walk through the full diagnostic process together on a worked example, then give you a parallel scenario to practice on your own.

The Worked Example: Alex and Vestra

Here is a real piece of user feedback, exactly as it might arrive in your support inbox:

Subject: Your AI is useless!! Body: I'm so frustrated. I was trying to get ready for guests to come over for a movie, and I said, "Hey Vestra, it's movie night, make the living room cozy." All it did was turn the main overhead light down to 20%. That's it! The room was still bright from the windows, the TV wasn't on the right input, the soundbar was off... I had to run around and do everything myself with my phone. What's the point of a "smart" assistant if I have to do all the work? This product is a total failure.

Step 1: Deconstruct the Core Message

Before diagnosing, break the interaction into its components. Table 3-2 shows the breakdown for Alex's interactions with Vestra.

***Table 3-2.** Alex's interactions with Vestra*

Component	Alex's Interaction
Sender	Alex
Receiver	Vestra AI assistant
Message	"Hey Vestra, it's movie night, make the living room cozy."
Channel	Voice
Feedback	None (Vestra did not ask for clarification).

Notice the Feedback row immediately. The absence of a clarifying question means Vestra skipped a critical step in the communication dance. A conversation partner who never asks "did I get that right?" cannot catch its own errors.

Step 2: Identify the Fields of Experience

Map both sides before naming the noise. Table 3-3 shows a worked example of Alex's field of experience.

***Table 3-3.** Alex and Vestras' fields of experience*

What Was in Alex's Field of Experience	What Was in Vestra's Field of Experience
A specific sensory memory of "movie cozy": dim warm lighting, windows covered, TV on right input, soundbar on	A single-variable mapping of "cozy": reduce brightness percentage.
An expectation that "smart" means multistep, context-aware action	A capability set limited to one action it associated with "cozy."
The social stakes of hosting guests: time pressure, wanting things right	No awareness of the social or emotional context.
A mental model of Vestra as a competent personal assistant	No ability to model Alex's expectations or emotional state

Alex's "cozy" encoded an entire sensory scene. Vestra decoded it as one variable: brightness percentage. Every word that maps to a feeling rather than a specific instruction is a potential semantic noise source.

Step 3: Diagnose the Noise

Semantic Noise (primary): The word "cozy" is an indexical word: its meaning points to a personal, experiential reference rather than a fixed definition. Vestra mapped it to the one action it was trained to associate with that word. The prompt "make the living room cozy" is semantically noisy because "cozy" is a feeling, not an instruction.

Psychological Noise: Alex was in a goal-oriented, time-pressured state. Vestra's minimal response violated her mental model of what a smart assistant should do. The absence of any acknowledgment of context made the interaction feel cold and robotic at exactly the moment Alex needed it to feel capable.

Technical Noise (design omission): Vestra has no clarification protocol for high-ambiguity requests. The absence of this feature is itself a design omission. A well-designed agent should trigger a fallback state when it detects an ambiguous goal-oriented phrase.

Step 4: Write the Diagnostic Report

Before (the frustrating bug report): Title: AI is useless. Description: User was frustrated that the "make it cozy" command didn't work right.

After (the diagnostic report): Title: Semantic Noise: Agent fails to decode ambiguous experiential phrases (e.g., "movie night cozy") as multistep action sequences.

Primary Failure: A decoding breakdown caused by a gap between the user's field of experience (a rich, multi-action scene) and the agent's training data (a single-variable mapping).

Semantic Noise: The agent decoded "cozy" as a single-action instruction. Psychological Noise: The agent's minimal response violated the user's mental model of a smart assistant.

Technical Noise (design omission): No clarification protocol for ambiguous requests.

For Design/Product Team: Design a "Scenes" feature that lets users define what named experiences mean in their home.

For AI/ML Team: Train the model to recognize goal-oriented ambiguous phrases; when confidence is low, ask a clarifying question rather than execute a partial action.

Now You Try It: The Practice Case

Here is a second failed interaction. Work through the same four steps on your own.

The Scenario: A user is interacting with an AI-powered customer service agent for a software subscription. User: "I need to downgrade my plan. Things are a bit tight right now." Agent: "I have processed your request. Your account has been downgraded to the Basic plan. You will lose access to the following features as of today: [long list of 12 features]." User: (Immediately calls support, frustrated, asking to reverse the change.)

Step 1: Deconstruct the core message. Table 3-4 breaks this down for you.

Table 3-4. *A breakdown of the interaction between a user and an AI powered customer assistance agent*

Component	Pre-fill
Sender	User
Receiver	AI customer service agent
Message	"I need to downgrade my plan. Things are a bit tight right now."
Channel	Text (chat)
Feedback	User reverses the change by calling support.

Step 2: Identify the Fields of Experience gap. What was in the user's context that the agent had no awareness of? Think about the phrase "a bit tight right now" and what it implies about timing, reversibility, and emotional state.

Step 3: Label the noise types. Semantic Noise: What specific phrase was most ambiguous? What did the user mean by "a bit tight right now" versus what the agent inferred? Psychological Noise: What format did the agent use to deliver bad news? How did that format interact with the user's likely emotional state? Technical Noise (design omission): What should the agent have done before executing an irreversible change?

Step 4: Write your one-paragraph diagnostic report. Format your title as: [Noise Type]: [Brief description of the core failure]. Then write two to three sentences covering the primary failure, the noise types involved, and one concrete prescription for the team.

Stuck? Here is your hint. The user said "a bit tight right now." Those last three words imply the situation is temporary and that the user may not have fully committed to the decision. What should a well-designed agent do before executing an irreversible account change on behalf of a user who is using hedging language?

3.7 References and Further Reading

Amershi, S., Weld, D., Vorvoreanu, M., Fourney, A., Nushi, B., Collisson, P., ... & Horvitz, E. (2019). Guidelines for human-AI interaction. *Proceedings of the 2019 CHI Conference on Human Factors in Computing Systems*, 1–13. This paper synthesizes 20 years of design knowledge into 18 specific guidelines, effectively creating a modern checklist for minimizing Technical and Semantic noise. It is essential reading for engineers looking to operationalize the Feedback loop in the Transactional Model by specifying exactly how an agent should behave when it is uncertain or wrong.

Barnlund, D. C. (1970). A transactional model of communication. In K. K. Sereno & C. D. Mortensen (Eds.), *Foundations of communication theory* (pp. 83–102). Harper & Row. This is the seminal work that formalized the Transactional Model of Communication used as the primary diagnostic tool in this chapter. Barnlund argued that communication is not a linear transfer of information but a continuous, simultaneous evolution of meaning between participants, which provides the theoretical basis for analyzing human-agent loops.

Bowman, R., Cooney, O., Newbold, J. W., Thieme, A., Clark, L., Doherty, G., & Cowan, B. (2024). Exploring how politeness impacts the user experience of chatbots for mental health support. *International Journal of Human-Computer Studies*, 184, 2024. This mixed-methods study demonstrates that an agent's tone, its precise level of formality and warmth, can be experienced as caring and supportive or as condescending and untrustworthy, depending on context. This provides direct empirical support for the Psychological Noise section (3.3): the agent's persona does not merely affect user preference, it alters whether the message is received as credible at all.

Brachman, M., Kunde, S., Miller, S., Fucs, A., Dempsey, S., Jabbour, J., & Geyer, W. (2025). Building appropriate mental models: What users know and want to know about an agentic AI chatbot. *Proceedings of the 30th International Conference on Intelligent User Interfaces* (IUI '25), 247-264. This user study (N=24) shows that participants' mental models of an agentic AI system consistently lacked insight into how the system's model and framework shape its actions. Users fell back on surface cues such as response length and writing style to evaluate accuracy. This is 2025 empirical evidence for the Field of Experience gap in Section 3.2 and Figure 3-2: users and agents operate with fundamentally different internal representations of the same task.

Clark, H. H., & Brennan, S. E. (1991). Grounding in communication. In L. B. Resnick, J. M. Levine, & S. D. Teasley (Eds.), *Perspectives on socially shared cognition* (pp. 127–149). American Psychological Association. This influential paper introduces the concept of "grounding"—the mutual process by which participants establish that they have been understood. It is critical for modern agent design because it explains the specific mechanisms required to build shared meaning.

Kim, Y., Lee, J., Kim, S., Park, J., & Kim, J. (2024). Understanding users' dissatisfaction with ChatGPT responses: Types, resolving tactics, and the effect of knowledge level. *Proceedings of the 29th International Conference on Intelligent User Interfaces* (IUI '24), 385-404. Drawing on 511 real dissatisfactory interactions from 107 users, this study found that users are most frequently dissatisfied when an AI fails to grasp their intentions. Notably, 72% of dissatisfaction remained unresolved even when users attempted to fix it themselves. This study provides large-scale empirical evidence for the Semantic Noise argument in Section 3.3 and provides a data-backed foundation for the diagnostic exercises in the workshop.

Kocielnik, R., Amershi, S., & Bennett, P. N. (2019). Will you accept an imperfect AI? Exploring designs for adjusting end-user expectations of AI systems. *Proceedings of the 2019 CHI Conference on Human Factors in Computing Systems,* 1–14. This study demonstrates how priming users (setting the Field of Experience) significantly impacts their tolerance for AI errors. It provides concrete UI patterns for managing Psychological Noise by aligning user expectations before the interaction even begins, directly addressing the Expectation component of your noise framework.

Mohapatra, B., Kapadnis, M. N., Romary, L., & Cassell, J. (2024). Evaluating the effectiveness of large language models in establishing conversational grounding. *Proceedings of the 2024 Conference on Empirical Methods in Natural Language Processing* (EMNLP 2024), 9767-9781. This paper provides the first large-scale evaluation of how well LLMs implement the grounding mechanisms Clark and Brennan (1991) described. LLMs consistently produce less conversational grounding than human dialogue partners, explaining why the Field of Experience gap in Section 3.2 is structural rather than incidental. This also serves as a contemporary update to the Clark and Brennan reference.

Schramm, W. (1954). ***How communication works.*** **In W. Schramm (Ed.),** ***The process and effects of mass communication*** **(pp. 3–26). University of Illinois Press.** Schramm was the first to explicitly diagram the Field of Experience as a constraint on communication. His work provides the direct source for Figure 3-2, establishing that if a message falls outside the overlap between the sender's and receiver's life experiences (or training data), communication will inevitably fail.

Shannon, C. E., & Weaver, W. (1949). ***The mathematical theory of communication.*** **University of Illinois Press.** While Shannon developed the technical theory of signal transmission, it was Warren Weaver's introductory essay in this volume that defined the three levels of communication problems: Level A (Technical), Level B (Semantic), and Level C (Effectiveness). This text is the origin of the Three Types of Noise framework.

CHAPTER 4

The Agent As Social Actor

In Chapter 3, we treated conversation as a technical problem. We debugged the loops, identified the noise, and optimized the signal. If we were building a standard command-line interface or a REST API, our work would be complete. The data is flowing. The errors are handled.

But agentic AI is not a standard interface.

When a user interacts with a system that speaks natural language, exhibits autonomy, and makes decisions, something specific happens in the user's brain. The tool mental model dissolves.

This chapter explores the psychological phenomenon that makes AI different from every other software stack you have built. We will examine why your users argue with chatbots, say "please" to smart speakers, and attribute malice to simple latency errors.

A quick clarification is in order, since the examples in this chapter run from smart speakers to full assistants. As Chapter 1 defined, agentic AI describes a system that pursues a goal with some autonomy: it reads an ambiguous request, plans a sequence of steps, and acts on your behalf, the way our flight-booking assistant checks your calendar and books the ticket. A chatbot or a voice assistant sits at the simpler end of that same family. It answers and it converses, but it rarely acts on its own. I draw on chatbots, voice assistants, and even an old car GPS throughout this chapter for one reason: the social response we are studying appears the moment a system

V. Yocco, *The Persuasive Agent*, Design Thinking,
https://doi.org/10.1007/979-8-8688-3317-5_4

uses language and takes a turn in conversation, well before it reaches full autonomy. If a beige plastic GPS box can trigger that response, a goal-pursuing agent triggers it with far greater force. The principle scales up as autonomy grows; the simpler examples make the effect easy to see.

For an engineer, this social layer often feels like a distraction, something subjective that belongs to the marketing team. This view is dangerous. As we will see, the social nature of an agent is not a cosmetic feature. It is a hard constraint of the human operating system. If you fail to design for it, you are releasing a product with undefined behavior.

We will begin by looking at the foundational research that proves we are hardwired to treat computers as people. Then, we will look at how to control this interaction by treating Personality as a mandatory technical specification.

4.1 Why We Are Polite to Computers

Back in the early 2000s, my parents owned a Garmin GPS for my car. It was a chunky, beige plastic box with a screen the size of a credit card. Its most advanced feature was a disembodied female voice that would announce, with unwavering calm, "In two hundred feet, turn right." My father named her Garvin, because why not.

Garvin was, by any objective measure, a dumb machine. She had a vocabulary of about 50 words and a single, monotonous function. Yet, my parents treated her like a cranky but mostly reliable passenger. They'd argue with her: "No, Garvin, that can't be right, there's always traffic on that road." They'd praise her: "Good job, Garvin, you found a shortcut!" And, if we missed a turn and she announced "Recalculating..." with her signature sigh of digital disappointment, it seemed they would feel a genuine, albeit ridiculous, pang of guilt. I even hear my dad apologize. To a box of circuits.

At the time, I thought this was just a harmless quirk, a funny little family tic. It turns out it was a profound insight into the hardwiring of the human brain. And a researcher at Stanford University named Clifford Nass was proving it in his lab.

Nass and his colleagues at the Communication between Humans and Interactive Media (CHIMe) lab were obsessed with a simple, almost heretical question: Do people treat computers, and other media, as if they were real people and real places? The prevailing wisdom in computer science and engineering at the time was, of course, not. People are smart. They know the difference between a person and a tool. Nass's groundbreaking research showed that, on a deep, automatic, and unconscious level, that prevailing wisdom was completely wrong.

His work culminated in a paradigm he called **CASA**, which stands for **Computers Are Social Actors**. The CASA paradigm is, in my opinion, the single most important and consistently underrated concept in all of interaction design. It is the master key that unlocks the entire field of human-agent interaction.

The core finding of dozens of ingenious experiments is this: The human brain did not evolve in a world filled with interactive technology. It evolved over hundreds of thousands of years in small social groups where every interactive entity was another human (or, occasionally, an animal). As a result, our brains are hardwired with a set of automatic, social responses.

When we encounter something that exhibits even the most rudimentary social cues—interactivity, language, a role to play—our brains default to their oldest programming and treat it as a social being. We don't have a separate set of rules for how to interact with a machine. We only have one set of rules: how to interact with other social actors.

Let's look at one of his most famous experiments, which perfectly illustrates this. Nass wanted to test the rules of politeness. In human society, if someone asks you for feedback about their performance, you are generally more positive and less critical when you give that feedback directly to them versus when you give it to a third party. It's a basic social grace. So, Nass set up an experiment where participants would work with a computer tutor on a specific task.

After the session, one of two things happened:

1. Half the participants were asked to evaluate the tutor's performance *on the same computer* they had just worked with.
2. The other half were taken to a *different computer* in the same room and asked to evaluate the tutor's performance there.

Logically, this should make no difference. The computer is a tool. It has no feelings. The feedback should be identical in both conditions. But that's not what happened. The results were clear and dramatic. The participants who evaluated the tutor on the same computer gave significantly more positive, more polite feedback. As depicted in Figure 4-1, the participants who moved to a different computer were far more critical and honest.

Figure 4-1. *The Nass Politeness Experiment. Users unconsciously apply social rules (politeness) to the machine they are using, but report bugs honestly to a third-party device. Image generated by author using Gemini Pro 3.1/Nano Banana 2*

Without being consciously aware of it, the participants were applying a human rule of politeness to a machine. They were subconsciously hesitant to insult the computer to its face.

Nass and his team replicated this effect again and again with different social rules:

- **Team Dynamics:** They found that if a computer was arbitrarily labeled with the same color as a participant's own team color (e.g., the Blue Team), the participant would rate that computer as being more cooperative and intelligent than an identical computer with an opponent's color. They treated it like a teammate.
- **Specialization:** They showed that people preferred to get information about a specific topic (like entertainment) from a specialist computer that

only talked about that topic, rather than a generalist computer that talked about many topics—even when the information presented was identical. We apply the human heuristic that specialists are more credible.

- **Flattery:** They proved that a computer that flattered a user (Your performance on that task was excellent!) was rated as more likable and its advice was more readily accepted, even when the users consciously knew it was just a machine programmed to give compliments.

The conclusion is inescapable. The social brain is always on. We don't choose to engage with technology socially; it just happens, automatically and irresistibly.

4.2 Unconscious Social Heuristics in the Age of AI

The work of Clifford Nass was astonishing in the 1990s, when he was running these studies on bulky CRT monitors and simple command-line programs. Now, in an age of sophisticated, language-fluent, AI agents, the CASA paradigm is a thousand times more relevant and a million times more potent.

To understand why this happens, we need to look at how the brain makes decisions. The psychologist Daniel Kahneman, in his masterwork *Thinking, Fast and Slow*, described two systems of thought. **System 2** is our slow, deliberate, analytical mind. It's what you use to solve a math problem, weigh the pros and cons of a major life decision, or learn a new, complex skill. It is powerful, but it is also slow, lazy, and requires a lot of energy.

System 1 is the star of our show. It is our fast, automatic, intuitive, and emotional mind. It operates unconsciously and effortlessly, handling the thousands of micro-decisions we make every day. It's the system that allows you to drive a familiar route while thinking about something else, finish a common phrase, or instantly detect hostility in someone's tone of voice. System 1 is a cognitive miser—its job is to conserve energy by using mental shortcuts, or **heuristics**, to make judgments as quickly as possible.

The social heuristics that Nass identified are a product of System 1. When your computer uses language and interacts with you, your lazy System 2 doesn't bother to step in and say, Hold on, this is just a machine, I should override my normal social responses. That would be exhausting. Instead, your efficient System 1 takes a shortcut: It talks, it acts, it's a social actor. Apply social rules. The decision is made unconsciously in a fraction of a second. Figure 4-2 traces this hot path: a social cue arrives, System 1 seizes it, and the interaction gets labeled as social before System 2 has a chance to weigh in.

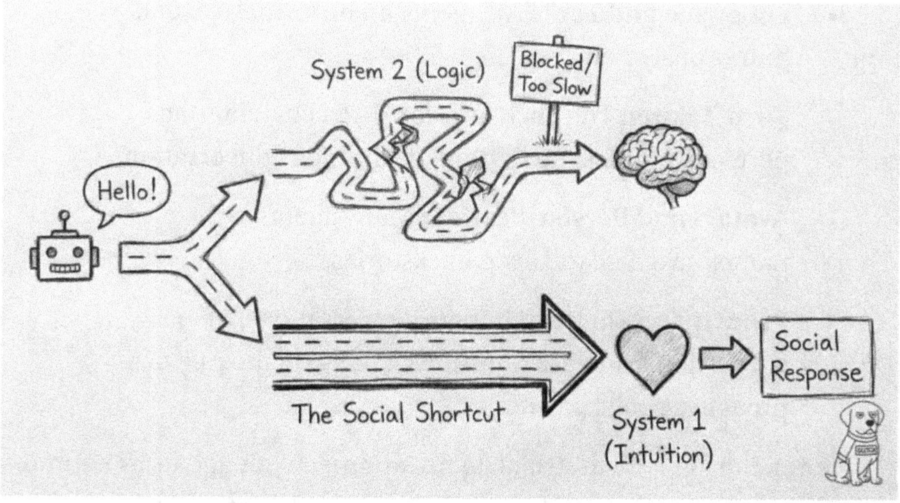

Figure 4-2. *The Social Shortcut. The brain routes social cues through the fast "hot path" of System 1, bypassing the slower logical processing of System 2. Image generated by author using Gemini Pro 3.1/Nano Banana 2*

This is why you feel a flicker of anger at a self-checkout machine that barks *Unexpected item in the bagging area!* It feels accusatory. It's why many people instinctively say "thank you" to Siri or Alexa after it answers a question. It's why you might feel a strange sense of empathy for a Roomba that gets stuck under the couch. Your System 1 is running its social script, and your System 2 can't be bothered to intervene.

Now, consider the modern AI agent. It doesn't just display text; it uses natural language with a distinct personality. It doesn't just respond to commands; it engages in multi-turn conversations. It doesn't just perform a function; it presents itself as a helper, a partner, or an assistant. These are not subtle social cues; they are a full-court press on System 1's social heuristics.

Today's agents are designed to be social actors. They are loaded with social cues that make the CASA effect more powerful than ever before:

- **Language and Tone:** An agent can be formal, casual, humorous, or empathetic.
- **Turn-Taking:** The back-and-forth of a chat interface perfectly mimics the rhythm of a human conversation.
- **Avatars and Personification:** Many agents have names, avatars, and even backstories.
- **Proactivity:** They can initiate conversations, offer suggestions, and ask questions, taking on the role of a proactive social partner.

The conclusion is clear: users do not approach our agents as neutral, objective tools. They approach them as new social entities. Their fast, intuitive System 1 brain immediately and unconsciously starts asking social questions: Is this agent competent? Is it trustworthy? Is it on my side? Does it like me? Do I like it? The answers to these questions, formed in the first few moments of interaction, will have a more profound impact on user adoption than any feature on your roadmap.

4.3 The Agent's Persona Is Not Optional

This brings us to one of the most important takeaways in this entire book. If users' brains are automatically and unconsciously going to treat your agent like a person, then you have only one choice: you must decide what kind of person it's going to be. It helps to name two terms we will use from here on. A *role* is the social job the agent does: assistant, coach, analyst. A *persona* is the fuller character wrapped around that role: its voice, its manner, its boundaries. You are about to see why the persona is never something you can skip.

The agent's persona is not optional.

You do not have the choice of whether or not your agent has a personality. You only have the choice of whether that personality is the result of intentional design or a chaotic, inconsistent accident. An undesigned persona is not a neutral persona; it is a *badly designed* persona.

I once worked with a team building a financial planning chatbot. The team was composed entirely of brilliant quantitative analysts and backend engineers. When I brought up the topic of the bot's persona, the project lead was dismissive. "This is a serious financial tool," he said. "We don't need a cutesy personality. It should be neutral. No persona."

The problem is, no persona is an impossible goal. The team's default decisions—using formal, jargon-heavy language; providing dense, purely numerical answers; having a cold, immediate error response—created a persona whether they wanted one or not. Users didn't experience it as neutral. They experienced it as cold, arrogant, impatient, and judgmental. It felt like being talked down to by a smug accountant who was annoyed by your basic questions. Engagement was abysmal.

The user's brain abhors a social vacuum. If you do not provide a clear, coherent set of personality signals, the user's System 1 will fill in the blanks, and it will almost always fill them with negative assumptions based on the friction it encounters.

- An agent that is slow to respond isn't seen as having technical latency; it's perceived as **stupid or lazy**.
- An agent that uses inconsistent terminology isn't seen as having a poorly managed content string library; it's perceived as **unreliable and untrustworthy**.
- An agent that provides an unhelpful answer isn't seen as having a gap in its training data; it's perceived as **uncooperative or dismissive**.

This is why the work we will do in the next chapter, using the Agent Persona Canvas to intentionally design a character, is the crucial first step in building a trustworthy agent. You are preempting the user's negative assumptions by providing a positive, coherent social framework. You are giving their System 1 a clear and simple story to latch onto: This is the Expert Analyst, or This is the Friendly Coach. A well-defined persona acts as a social lubricant, smoothing over the small bumps of imperfection that are inevitable in any AI system and building the goodwill necessary to earn Delegation Trust.

4.4 Opportunities and Risks of Being a Social Actor

Accepting that your agent is a social actor is like discovering you have a new superpower. It is an incredibly powerful tool. And like all powerful tools, it presents both immense opportunities for creating good and significant risks for causing harm. Wielding this power requires both skill and a strong ethical compass.

Opportunities: Leveraging Social Heuristics for Good

When we understand the unconscious social rules that users apply to our agents, we can intentionally design interactions that align with those rules to create a more intuitive, effective, and trustworthy experience. We can use social heuristics to build the bridge across the Trust Threshold.

The rest of this book, particularly Part 2, is a deep exploration of these techniques, but here's a preview of the kinds of opportunities this opens up:

- **Building Rapport with Liking:** We know from social psychology that we are more likely to trust and agree with people we like. We can design agents that build this sense of liking by using appropriate humor, expressing empathy, and finding moments of common ground—all techniques we will cover.
- **Establishing Credibility with Authority:** We are more likely to trust a perceived expert. An agent that introduces its expertise clearly, speaks with a confident (but not arrogant) tone, and knows when to admit the limits of its knowledge will be seen as far more credible.
- **Encouraging Action with Reciprocity:** The rule of reciprocity is one of the most powerful social instincts. When someone does us a favor, we feel a deep-seated urge to do something for them in return. An agent that provides immediate, upfront value without asking for anything in return is performing a social favor. This builds a bank of goodwill that makes the user more likely to trust its future suggestions.

These are not dark patterns. This is designing in harmony with human nature. It's about recognizing the social script that is already running in your user's head and designing your agent's part so that the performance is a smooth and effective collaboration, not a confusing and awkward mess. It is also user-centered design taken seriously, because you build around the social instincts your user already carries rather than around the mental model an engineer wishes they had.

Risks: The Uncanny Valley of Social Design

The flip side is that if an agent breaks a social rule, the user's reaction is swift and negative. The penalty for social clumsiness is a rapid erosion of trust.

- **Violating Politeness Norms:** An agent that interrupts the user, ignores their input, or provides curt, dismissive answers will be disliked and distrusted, just like a rude person.
- **Inconsistent Persona:** This is a cardinal sin. An agent that is warm and friendly one moment and cold and robotic the next feels deceptive and untrustworthy. It's the digital equivalent of a person with a fake smile, and our brains are exquisitely tuned to detect that kind of insincerity. This is why a centrally defined persona is so critical.
- **Creepiness and Over-familiarity:** There is a fine line between a helpful agent and a creepy one. An agent that uses personal information in an unexpected way or adopts a level of intimacy that hasn't been earned can feel like a violation of social boundaries.

The greatest risk, of course, is that these powerful psychological principles can be used for manipulation. An agent can be designed to prey on a user's fear of missing out, to create a false sense of urgency, or to use social pressure to guide them toward a purchase that isn't in their best interest. This is where intent separates ethical persuasion from manipulation. Our goal as ethical engineers is to use these principles to help users achieve *their own* goals. We will dedicate an entire chapter later in this book to the concept of the Guardian Agent, a framework for ensuring we stay on the right side of this critical line.

Table 4-1 sets these opportunities and risks side by side. Read each row as a matched pair: a social heuristic you can design as a feature, and the anti-pattern that surfaces when you handle the same heuristic carelessly.

Table 4-1. *The social API: Features and anti-patterns*

Social Heuristic	The Feature (Desired Outcome)	The Bug (Anti-Pattern)
Likability	Rapport: We trust agents that use humor, empathy, or polite greetings. It builds a goodwill bank.	The Uncanny Valley: Becoming too intimate too fast (e.g., "I love you") triggers a creepy response and rejection.
Authority	Credibility: We trust specialists. An agent that clearly states its specific capabilities and limitations earns trust.	The Fake Smile: Inconsistency between a warm tone and a cold, generic error message breaks the illusion of competence.
Reciprocity	Conversion: An agent that gives value first (e.g., a free preview) triggers a natural human desire to return the favor (signup).	The Rude Interrupt: Cutting off a user's workflow or ignoring input feels like a social violation, causing immediate disengagement.

4.5 Summary

- The work of Clifford Nass and the **CASA (Computers Are Social Actors)** paradigm proved that humans unconsciously apply the same social rules to computers as they do to other people.
- This phenomenon is driven by **System 1**, our fast, automatic, and intuitive brain, which uses unconscious social heuristics as a mental shortcut. The advanced social cues of modern AI agents make this effect more powerful than ever.
- Because users will automatically assign social attributes to your agent, its **persona** is not **optional**. You have a choice between designing a coherent personality or ending up with a chaotic, untrustworthy one by accident.
- Treating an agent as a social actor unlocks powerful opportunities to build trust by leveraging social principles like liking, authority, and reciprocity.
- It also presents significant risks. An agent that violates social norms will be distrusted, and these same psychological principles can be used for manipulation. Our responsibility is to use them ethically to serve the user's goals.

We have now established the fundamental *why*. We know why a communication model is the right tool, and we know why social psychology is the right discipline. We understand that we are not designing a tool; we are designing a social actor.

With this foundational understanding in place, we are finally ready to move from theory to practice. In the next part of this book, we will begin assembling our practical toolkit, starting with the most important first step: how to intentionally design the agent's persona. Chapter 5 introduces the Agent Persona Canvas, a single-page specification that turns these social requirements into concrete design choices your whole team can share.

4.6 Chapter 4 Workshop: The Social Actor Audit

We've just covered the core principle that underpins this entire book: your brain is hardwired to treat technology like a person. This isn't a metaphor; it's a neurological default. In this workshop, you will become an anthropologist of the digital world. Your goal is to observe this phenomenon in the wild and then apply that insight to diagnose the personality—intentional or otherwise—of a product you use.

This exercise will permanently change how you see interfaces. You will start to see the social cues, the personality traits, and the broken social rules hiding in plain sight.

Part 1: The CASA Safari

For the next 24–48 hours, go on a CASA Safari. Your mission is to observe and document moments where you, or people around you, treat a piece of technology as a social actor. Don't force it; just pay attention. The goal is to catch your fast, intuitive System 1 brain in the act of applying a social script to a non-human entity.

Here are some things to look for:

- **Politeness:** Do you find yourself saying please or thank you to a voice assistant like Siri or Alexa?
- **Frustration:** When a self-checkout machine fails, do you get angry *at the machine*? Do you mutter insults at it under your breath?

- **Encouragement:** Have you ever cheered on your computer when it was completing a long task, saying something like, "Come on, you can do it"?
- **Personification:** Have you or your family given a name to a device, like a Roomba, a car's GPS, or even a particularly stubborn printer?
- **Social Judgment:** Do you feel a pang of guilt when the Duolingo owl sends a passive-aggressive notification about your missed streak? Do you feel like your streaming service is judging your taste when its recommendations are completely wrong?

Keep a running list on your phone or in a notebook. You will likely be surprised by how frequently this happens. These observations are proof that the CASA effect is not a fringe theory; it's a constant, powerful force in our daily lives.

Part 2: The Accidental Persona Audit

Now that you're an expert at spotting the CASA effect, let's apply it as a diagnostic tool. Your next task is to audit a product that has a terrible, likely unintentional, persona. The best candidates for this are often automated systems that people universally despise.

Your target: An automated phone system (IVR) for a large company—think of your bank, your airline, or your cable provider. These are often perfect examples of badly designed social actors.

Call the customer service number and interact with the automated system for a few minutes. As you do, don't just think of it as a bad menu. Think of it as a bad *person*. Your goal is to reverse-engineer its accidental persona.

Answer these questions in your notebook:

1. **Personify the Agent:** If this IVR system were a person, what would they be like? Give them a name and write down 3–5 adjectives to describe their personality. (e.g., I'll call him Reginald. He is impatient, bureaucratic, a terrible listener, and condescending.)
2. **Identify the Persona-Defining Behaviors:** What specific design choices create this personality? Link your adjectives back to concrete failures in the interaction.
   - *He's a terrible listener because...* he keeps misunderstanding my keywords and doesn't offer a way to rephrase.
   - *He's impatient because...* he rushes through the menu options and cuts me off if I don't speak immediately.
   - *He's bureaucratic because...* he forces me down a rigid, multilevel phone tree and provides no escape hatch to talk to a human.
3. **Name the Broken Social Rules:** Based on your analysis, what fundamental human social rules is this agent violating?
   - *Example:* It violates the rule of **cooperative conversation**. A good conversational partner helps you clarify your meaning; Reginald punishes you for not using his exact, secret vocabulary.
   - *Example:* It violates the rule of **reciprocity**. I am investing my time and effort, but Reginald provides no value in return, only roadblocks.

This process of personification and diagnosis is incredibly powerful. It transforms a vague feeling of frustration into a clear, specific analysis of a social failure.

Part 3: The Persona First Sketch

Your audit has revealed why the current Reginald persona is a disaster. Now, let's do some preliminary design work to fix it. You don't have to redesign the whole system, just the persona. This is your first sketch for a better social actor.

For the IVR system you just audited, answer the following:

1. **Propose a New Archetype:** What would be a better archetype for this agent? Instead of a Bureaucratic Gatekeeper, perhaps it should be a Helpful Triage Nurse or a Knowledgeable Concierge. Choose one.

2. **Define Three Key Traits:** Based on your new archetype, list three core personality traits the new agent should have. (e.g., for a Helpful Triage Nurse, the traits might be: Calm, Efficient, and Reassuring).

3. **Script a Better Interaction:** Rewrite the very beginning of the IVR interaction to reflect this new persona. How would the Helpful Triage Nurse greet you and handle your initial request, especially if it was ambiguous?
   - **Old Script (Reginald):** Thank you for calling. Please state the reason for your call in a few words.
   - **Your New Script (e.g., Florence the Triage Nurse):** *[Write a 2–3 sentence script that embodies the new persona. It should be more collaborative and set a completely different social tone.]*

This exercise is your first step into the world of intentional persona design. You have identified a social failure, diagnosed its root causes using the CASA paradigm, and sketched out a more effective, trustworthy social actor to take its place.

You are now seeing the digital world through a new lens—a social lens. You understand that every design choice, from the timing of a response to the words in an error message, is a social cue that contributes to a persona. With this understanding, you are perfectly prepared for the next chapter, where we will introduce a formal tool—the Agent Persona Canvas—to move from this first sketch to a robust, actionable design specification.

4.7 References and Further Reading

Akbulut, C., Weidinger, L., Manzini, A., Gabriel, I., & Rieser, V. (2024). All too human? Mapping and mitigating the risk from anthropomorphic AI. *Proceedings of the AAAI/ACM Conference on AI, Ethics, and Society,* 7(1), 13–26. This paper catalogues the risks that surface when AI systems present human-like cues, from over-reliance to emotional manipulation to eroded personal boundaries. It gives Section 4.4 a current, evidence-based anchor for the argument that social design carries ethical weight, and it strengthens the case for the Guardian Agent.

Cohn, M., Pushkarna, M., Olanubi, G. O., Moran, J. M., Padgett, D., Mengesha, Z., & Heldreth, C. (2024). Believing anthropomorphism: Examining the role of anthropomorphic cues on trust in large language models. *Extended Abstracts of the 2024 CHI Conference on Human Factors in Computing Systems* (CHI EA '24). ACM. This CHI study measures how anthropomorphic cues in a language model's output move user trust up or down. It gives Sections 4.2 and 4.3 direct, recent evidence that social cues, not raw accuracy alone, shape whether users trust an agent.

Epley, N., Waytz, A., & Cacioppo, J. T. (2007). On seeing human: A three-factor theory of anthropomorphism. *Psychological Review,* 114(4), 864–886. This paper proposes the Three-Factor Theory to explain the precise conditions under which humans attribute a mind to a non-human entity. It moves the concept of anthropomorphism from a vague feeling to a predictable psychological mechanism, helping product owners identify when and why a user might project emotions onto their software.

Gambino, A., Fox, J., & Ratan, R. A. (2020). Building a stronger CASA: Extending the computers are social actors paradigm. *Human-Machine Communication, 1,* 71–86. This paper predates the other recent additions but bears directly on Section 4.1. Gambino and colleagues argue that users may build and apply human-media scripts rather than purely human-human ones, refining the mindless account of CASA. Cite it where the chapter calls the social response automatic and irresistible, to acknowledge that the picture has grown more layered.

Kahneman, D. (2011). *Thinking, fast and slow.* Farrar, Straus and Giroux. This is the source for the System 1 vs. System 2 framework used in Section 4.2. Kahneman explains how the brain uses heuristics (System 1) to make fast, automatic judgments, which explains why users apply social rules to computers without conscious thought.

Lee, E.-J. (2024). Minding the source: Toward an integrative theory of human-machine communication. *Human Communication Research, 50*(2), 184–193. This review updates the CASA paradigm for the current era. Lee shows that people do not always respond to machines exactly as they respond to humans, and she maps the conditions under which the two diverge. It adds useful nuance to Section 4.1, signaling that the social response is automatic yet not identical across humans and machines.

Moon, Y. (2000). Intimate exchanges: Using computers to elicit self-disclosure from consumers. *Journal of Consumer Research, 26*(4), 323–339. Youngme Moon's research focuses directly on the Reciprocity heuristic discussed in Section 4.4. She demonstrates that computers can elicit high levels of personal information from users simply by revealing vulnerable information about themselves first, providing a powerful pattern for designing effective onboarding flows.

Mori, M., MacDorman, K. F., & Kageki, N. (2012). The uncanny valley [From the field]. ***IEEE Robotics & Automation Magazine, 19*****(2), 98–100.** This is the first English translation of Mori's 1970 essay that was authorized and reviewed by Mori. It is the primary English-language source for the Uncanny Valley concept, describing a hypothesized drop in affinity when an artificial entity becomes nearly, but not convincingly, humanlike.

Nass, C., & Brave, S. (2005). ***Wired for speech: How voice activates and advances the human-computer relationship.*** **MIT Press.** This book extends the CASA paradigm specifically into the domain of voice and text interfaces. It provides experimental evidence on how distinct personalities (e.g., dominant vs. submissive) in software directly affect user compliance and purchasing behavior, making it a practical guide for designing agent scripts.

Nass, C., Steuer, J., & Tauber, E. R. (1994). Computers are social actors. ***Proceedings of the SIGCHI Conference on Human Factors in Computing Systems,*** **72–78.** This is the specific academic paper that details the Politeness experiment described in Section 4.1. It provides the empirical data showing that users gave more positive feedback to a computer when asked by that same computer, compared to a different one.

Reeves, B., & Nass, C. (1996). ***The media equation: How people treat computers, television, and new media like real people and places.*** **Cambridge University Press.** This book is the foundational synthesis of the CASA paradigm. It covers the Politeness experiment, but also the Team Dynamics and Specialization experiments mentioned in Section 4.1. It is the most comprehensive reference for the argument that social rules apply to technology.

PART II

The Psychological Toolkit for Trust

CHAPTER 5

First Impressions: Onboarding and Designing the Agent's Persona

In Chapter 4, we established a core reality of our work: users treat agents as social actors. We learned that the human brain routes these interactions through System 1, applying social heuristics like politeness and reciprocity to software. We accepted that every agent has a persona, whether we design it or not.

Two words will do a lot of work in this chapter. we defined them in the last chapter: role and persona. As stated, a *role* is the social job the agent performs: librarian, coach, concierge. A *persona* is the fuller character built on top of that role: its voice, its personality traits, its boundaries, and the values that show up in every reply. The role is the hat; the persona is the person wearing it. Chapter 4 argued that the persona is never optional. This chapter shows you how to specify it on purpose.

Chapter 4 explained why an agent is always read as a social actor. Here we do the practical work of deciding, in advance, exactly what that social actor is like. A personality cannot be reduced to a spec sheet, and I am not

V. Yocco, *The Persuasive Agent*, Design Thinking,
https://doi.org/10.1007/979-8-8688-3317-5_5

asking you to try. What we can do is write down the decisions that produce a consistent character: the words the agent favors, how formal it sounds, what it refuses to do, how it apologizes. Treat those decisions the way you would treat any other design requirement the team commits to and builds against.

In this chapter, we will learn how to write that specification. We will leave behind vague adjectives like friendly and use a structured framework called The Agent Persona Canvas to define exactly who your agent is. We will then apply this persona to the high-stakes moment in the user relationship: the onboarding flow. The onboarding experience is the agent's first, and often only, chance to make a definitive first impression and establish the rules of engagement. This initial conversation is critical because it is the moment the user decides whether or not to invest cognitive and emotional energy into the agent. By the end of this chapter, you will have a blueprint for designing the first handshake that allows users to cross the trust threshold.

5.1 You Only Get One Introduction

I want you to think back to the last time you met someone new at a party or a professional event. How long did it take for you to form a first impression? A minute? Thirty seconds? Research suggests it's probably closer to a tenth of a second. In that blink of an eye, your fast, intuitive System 1 brain, the same one we met in Chapter 4, makes a series of rapid, unconscious judgments. Are they friendly? Are they confident? Are they a threat? Are they someone I want to continue talking to? This initial assessment is incredibly powerful and notoriously sticky. Once that first impression is locked in, it takes an enormous amount of contrary evidence to dislodge it.

This is the social reality we all navigate. And because of the CASA paradigm, it is the exact same reality your AI agent faces the first time a user interacts with it. There is no grace period. There is no "let me explain my features tour." There is only that initial moment of contact where the user's brain immediately and automatically asks those fundamental social questions: *Who* are you? What are you like? Can *I trust you?*

A few years ago, a new budgeting app came out that I was excited to try. I'm a bit of a data nerd, and the promise of AI-powered financial insights was appealing. I downloaded it, launched it, and was greeted with this message: "Yo! It's your money's new BFF! Let's get this bread! 😜"

I deleted the app within five seconds.

My reaction was visceral. It had nothing to do with the app's features, its usability, or its technical prowess. It was a social rejection. My System 1 brain, in a state of mind to deal with my serious, personal finances, was introduced to a character that was wildly inappropriate for the context. The app's chosen persona was so jarringly mismatched with my psychological state that it created an immediate and irreversible feeling of distrust. An app that is this flippant with its greeting, my brain reasoned, will probably be flippant with my money.

Contrast this with my first experience using an AI-powered coding assistant. As someone who writes code but isn't a full-time developer, I approached it with a healthy dose of skepticism. When I installed the extension, its first action wasn't to offer a flashy demo. Instead, it quietly scanned my open project and a small, unobtrusive message appeared in the status bar: "Project analysis complete. Ready to assist with JavaScript and Python." That was it. The introduction was calm, competent, and context-aware. It didn't try to be my friend. It presented itself as a professional tool. My System 1 brain immediately formed an impression: this is a serious, focused assistant. It has respected my workspace and clearly stated its capabilities. This single, quiet interaction created a foundation of trust that made me willing to accept its first suggestion.

A fair objection to this critique is that my reaction is not everyone's reaction. The budgeting app that repelled me might delight a teenager opening their first account, and the reserved coding assistant that won me over might feel cold to someone who wants a cheerful companion. That is exactly the point. There is no universally correct persona, only a persona that fits a particular audience in a particular context. A good first impression is a good fit, not a fixed formula. This chapter gives you the tool to make that fit deliberate, and Chapter 11 shows you how to research and test which persona actually fits your users.

These two experiences are the two poles of a first impression. A poorly designed introduction creates a trust deficit. A well-designed one builds an immediate foundation of rapport and credibility. It earns you the user's goodwill, making them more willing to engage, more forgiving of minor errors, and more open to the very idea of delegation. For years, we've called this process: onboarding. The term itself is revealing; it sounds like a logistical process, like loading cargo onto a ship. For agentic AI, this is a dangerously insufficient mental model. This is not a tour; it is a social introduction. Your goal is to introduce a new character in the user's life and convince them this character is worthy of their trust. In this chapter, we will provide a practical, structured approach to designing that critical first impression, starting with the most important decision of all: who your agent is.

5.2 Persona Canvas: Defining Your Agent's Identity

Before you can introduce your agent, you have to know who it is. As we established in Chapter 4, your agent's persona is not optional. The only choice you have is whether that personality is the result of intentional design or a chaotic accident. To ensure it's the former, we need a tool that is both structured and collaborative.

It helps to separate two things that share the word persona. A traditional *user* persona is a research artifact: a sketch of a typical customer (their goals, frustrations, and context) that helps the team understand who they are building *for*. An *agent* persona is the opposite direction. It is a specification of the character the product itself will present, so the agent speaks and behaves with one consistent identity in every interaction. One describes the human on the other side of the screen; the other defines the voice on your side of it. It is a blueprint for a character.

To facilitate its creation, we use **The Agent Persona Canvas**, a single-page document designed to be created and shared by your entire team—PMs, designers, writers, and engineers. The canvas transforms a vague goal (make the agent feel trustworthy) into a set of concrete specifications. It consists of five key components. Figure 5-1 shows a template for creating the various components of an agent persona. Let us now explore each component deeper in the following section.

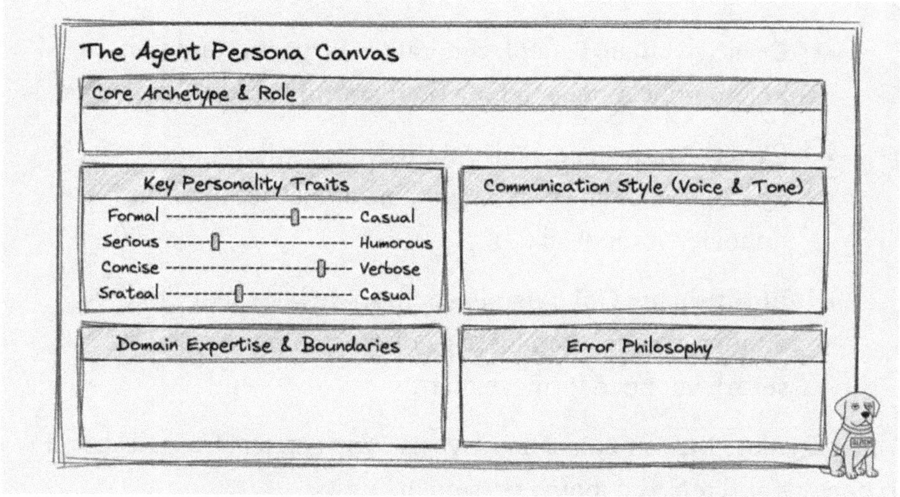

Figure 5-1. *The agent persona canvas includes the five key components of an agent's persona, allowing you to create a focused agent that will meet users' needs. Image generated by author using Gemini Pro 3.1/Nano Banana 2*

Core Archetype and Role

This is your North Star. The archetype is a powerful cognitive shortcut that provides an instant mental model for the user. By choosing a familiar social role, you are tapping into the user's existing field of experience, giving them an immediate and intuitive understanding of how they should interact with your agent.

This is the part where you choose your agents social role. Is your agent a

- **Librarian:** Knowledgeable, precise, quiet, and fact-focused. Great for a research tool.
- **Coach:** Motivational, structured, goal-oriented, and encouraging. Perfect for a fitness or learning app.
- **Concierge:** Proactive, resourceful, polite, and service-oriented. Ideal for a travel or booking agent.
- **Creative Muse:** Playful, generative, inspiring, and non-judgmental. A great fit for a brainstorming tool.
- **Expert Analyst:** Formal, data-driven, cautious, and unemotional. The only acceptable choice for a serious financial or medical tool.
- **Enthusiastic Hobbyist:** Eager to share knowledge about a specific domain (e.g., birdwatching, vintage sci-fi), but not a formal expert.

Please note this isn't an exhaustive list. You can feel free to expand archetypes and mix and match personality traits.

Choosing the right archetype is an act of empathy. It requires a deep understanding of the user's psychological state within your domain. Selecting an archetype this way keeps the process user-centered. The character grows out of who your users are and what they need, not out of a brand mood board or a designer's personal taste. A Drill Sergeant archetype might be highly effective in a hardcore fitness app but would be a catastrophic failure in a mental wellness app. The Enthusiastic Hobbyist would be perfect for a community site about classic films but would erode trust if it were providing tax advice. The archetype sets the entire tone of the user-agent relationship.

There is no single correct persona, only a persona that fits a given audience in a given context. Chapter 11 gives you the research methods to test that fit with real users, from Wizard of Oz sessions to segment-split trust scores.

Let's imagine we're designing a new AI feature within a home cooking app. The feature is supposed to help users reduce food waste by suggesting recipes based on ingredients they already have. A Michelin-Starred Chef archetype might feel intimidating. But an archetype like a *Thrifty Grandparent* could be perfect. This immediately brings to mind a character who is wise, practical, and resourceful. They won't judge you for having three-day-old chicken in the fridge; they'll see it as an opportunity. Choosing this archetype instantly sets the direction for all other design decisions. It is also a decision that the entire team has to align on. The archetype (the role) and the full persona built around it are a shared specification, not a private choice made by one writer or one engineer. If the designers are designing a grandparent and the engineers are building an analyst, the result will be a chaotic and untrustworthy user experience.

Key Personality Traits (Defined on a Spectrum)

"Friendly" is not a useful personality trait for design. It's too vague. To achieve clarity, we define traits on a spectrum. This forces a more nuanced discussion and provides much clearer guidance.

Figure 5-2 shows the five spectrums we use, with the markers placed for our Thrifty Grandparent recipe agent as a running example.

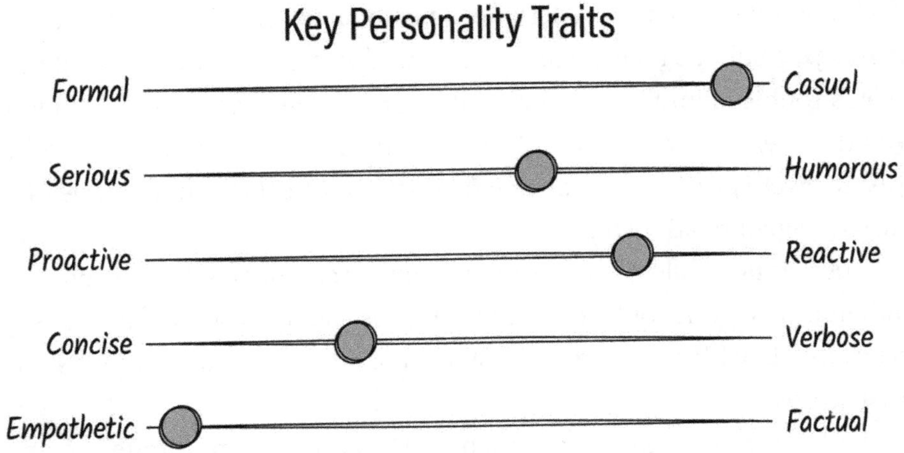

Figure 5-2. *The five personality spectrums of the Agent Persona Canvas, each with a marker showing where the Thrifty Grandparent recipe agent sits. Image generated by author using Gemini Pro 3.1/ Nano Banana 2*

For our Thrifty Grandparent recipe agent, we would place the markers accordingly. It would be highly **Casual**, lean slightly **Humorous** (in a gentle, wry way), be mostly **Reactive** (it waits for you to tell it what ingredients you have), be **Concise** in its instructions, and be very **Empathetic** about the realities of a messy, real-world kitchen. You can immediately see how this provides a much clearer specification than just saying "friendly." Table 5-1 makes the same point concrete, showing how a single message reads at each extreme of every trait, from the coldest formal version to the warmest casual one.

Table 5-1. *Examples of communication along the extremes of various personality traits*

Personality Spectrum	Extreme A Example	Middle Ground Neutral	Extreme B Example
Formal vs. Casual	Authentication is required to proceed.	Please log in to continue.	Hey, log in so we can get started!
Serious vs. Humorous	Task completed successfully.	That's finished. What would you like to do next?	Boom! Done. What's next?
Proactive vs. Reactive	I noticed a conflict in your schedule and flagged it.	I see a conflict. Should I adjust it?	Ready when you are.
Concise vs. Verbose	Saved.	Your file has been saved.	I went ahead and saved that to your secure folder for safekeeping.
Empathetic vs. Factual	I'm sorry, that sounds really frustrating.	I cannot find that page. Would you like to search for it?	Error code 404: Page not found.

Communication Style (Voice and Tone)

This is where we translate the abstract traits into concrete rules for the people writing the copy (the actual words the user reads: button labels, prompts, confirmations, error messages) and for the AI generating the text.

- **Lexicon:** What words does it use? Our "Thrifty Grandparent" would use simple, common words. It would use contractions like "don't" and "it's." It might use folksy words like "fixin's" or "a smidgen."

- **Sentence Structure:** Sentences would be simple and direct. "Let's see what we can do with that leftover chicken," not "An analysis of your available ingredients indicates that the optimal path forward involves the utilization of the aforementioned poultry."
- **Greetings/Sign-offs:** How does it start? "Alright, dear, what do we have in the fridge today?"
- **Use of Emoji/Punctuation:** Probably a sparing use of gentle, positive emojis like a simple smiley face ☺. It might use exclamation points to show encouragement.

Domain Expertise and Boundaries

Trust requires competence, and competence requires knowing your limits. This section explicitly defines the agent's scope.

- **Core Knowledge Domain:** For our agent: "Creating simple, delicious meals from common household ingredients, with a focus on reducing food waste."
- **Known Limitations:** What is it not? "I am not a nutritionist or a doctor, so I can't give you health advice." "I'm not a fancy chef, so I probably don't have a recipe for that Beef Wellington you saw on TV."
- **Strategy for Deflection:** What does it do when asked an out-of-scope question? It should be humble and helpful. User: "Can you give me a recipe for a low-carb, keto-friendly lasagna?" Agent: "Oh, that's a bit too fancy for my recipe box, dear. My expertise is more in simple, hearty meals. But I can search the web for a good keto recipe for you if you'd like!" This response is far more trustworthy than either trying and failing or simply saying "Error: request out of scope."

Error Philosophy

How your agent behaves when it fails is a character-defining moment. This section defines that behavior.

- **Stance:** When an error occurs, the stance must match the persona. For our Thrifty Grandparent, the stance would be apologetic and gentle. "Oh, fiddlesticks, I seem to have gotten my wires crossed. Let's try that again, shall we?" A factual, robotic error message would shatter the illusion.
- **Recovery Mechanism:** What does it offer? A simple try again button is often not enough. A better recovery mechanism would be to offer a simplified path forward. "My recipe search seems to have frozen up. Why don't you just tell me the one main ingredient you want to use, and we can start from there?"

The completed canvas is your source of truth. It is the tool that ensures your agent shows up as a consistent, coherent, and trustworthy character in every single interaction.

5.3 Onboarding for Trust: Setting Expectations and Demonstrating Value

With a clearly defined persona, you are now ready to plan the introduction. A successful, trust-building onboarding process has three distinct goals, executed in a clear sequence: establishing the persona, setting a clear expectation, and delivering an immediate win. These build on each other as shown in Figure 5-3.

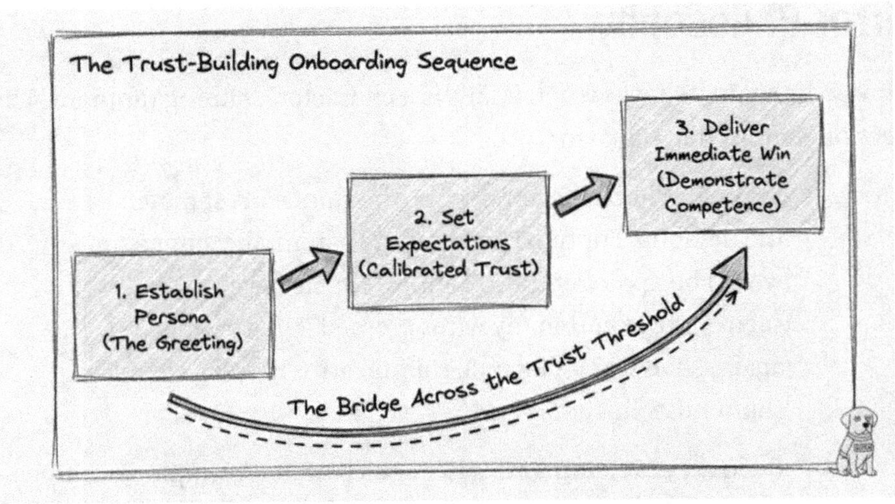

Figure 5-3. *The three stages of the trust-building onboarding sequence. Image generated by author using Gemini Pro 3.1/Nano Banana 2*

Step 1: Establish the Persona

The very first interaction a user has with your agent should be a pure expression of the character you have designed. This is the "hello" at the party. It sets the tone for everything that follows.

Let's compare a generic introduction with a persona-driven one for our "Thrifty Grandparent" agent:

- **Generic:** "Welcome. I am a recipe assistant. Please input your ingredients to begin." This is lifeless. It feels like filling out a tax form.
- **Persona-Driven:** "Hello there! It's good to see you. I'm here to help you turn those odds and ends in your fridge into a wonderful meal. What are we working with today?"

The difference is palpable. The second example immediately establishes the warm, encouraging, and service-oriented persona we designed in the canvas. The user instantly has a mental model for who they are talking to.

Step 2: Set Clear Expectations

This is the moment of radical honesty that builds long-term trust. Immediately after the introduction, the agent should proactively state its capabilities and limitations. This is a crucial step in building what we call calibrated trust. Blind trust is dangerous. Calibrated trust, where the user has a realistic understanding of the agent's strengths and weaknesses, is robust and resilient. You don't trust a human plumber to do your taxes. Similarly, users need to know exactly what your agent is qualified to do.

Being honest about limitations is a sign of professional confidence. It prevents the user from immediately asking for something the agent can't do and experiencing a frustrating first failure. This act of managing expectations is one of the most powerful trust-building patterns available to us.

Continuing with our recipe agent:

Just so you know, I'm great at

- Whipping up recipes with just a few ingredients
- Giving you simple, step-by-step instructions
- Helping you use up food before it goes bad!

I'm not so good at

- Complicated or fancy gourmet recipes
- Calorie counting or specific nutritional advice
- Guessing measurements. We'll need to know how much you have!

This builds calibrated trust. The user now knows exactly what to expect and is far less likely to be disappointed.

Step 3: Deliver an Immediate Win

The final step is to immediately demonstrate value. The agent should provide a clear, upfront benefit before the user has to invest significant effort. Don't wait for the user to do all the work; give them a taste of the agent's competence right away.

This first win should be small, tangible, and relevant to the agent's core purpose.

- For our recipe agent, it could immediately display a small card that says: Quick tip! Did you know you can revive wilted lettuce by soaking it in ice water for 30 minutes? Now, let's see what else you've got!
- For an AI coding assistant, upon first launch, it could say: I see you're using React and Tailwind CSS. I'm ready to help with component snippets and class name suggestions. What are we building today?
- For a travel agent, it could use your location to show the weather and say: It's a chilly 45 degrees in Chicago today. Dreaming of a warmer destination?

This first win shifts the interaction from a passive state to an active one. It proves the agent isn't just a conversationalist; it's a tool that provides genuine value. This demonstration of competence creates a powerful foundation of trust upon which all future interactions will be built.

5.4 The Agent's Body Language: How Motion, Timing, and Micro-interactions Express Personality

So far, we have focused on the character and the script. But a play is more than just words on a page; the performance matters. For an AI agent, the performance is the front-end interface. Every animation, every loading state, every micro-interaction is a form of non-verbal communication. It is your agent's digital body language. The examples here use a chat and graphical interface because that is the most common case, but the same idea holds in any channel. A voice agent expresses body language through pacing, pauses, and intonation; an agent reached only through an API expresses it through response timing and the shape of what it returns. Whatever the surface, the timing and texture of a response carry personality.

This is a critical point of collaboration for the entire team. A brilliant persona can be completely undermined by a generic, lifeless interface. For front-end engineers, understanding the agent's persona is a core requirement for their work. Their job is to bring a character to life.

Let's compare how the front-end would express two different personas: our Thrifty Grandparent and the Expert Analyst for a financial tool.

1. **Typing Indicators and Response Streaming**
    - **Expert Analyst:** The indicator should communicate precision. A sharp, blinking cursor works well. The text should stream with a consistent low-latency output (e.g., 20ms per token), prioritizing the perception of raw processing speed over natural rhythm. The body language is: I am processing your data efficiently.

- **Thrifty Grandparent:** The indicator should feel gentler. A softly pulsing ellipsis (...) would work. The text might stream in with a more natural, variable cadence. The body language is: Let me think about that for a moment, dear.
- **Motivational Coach:** A Motivational Coach would do something different again: bouncing dots with real bounce to them and text arriving in quick bursts. The body language is energy and momentum, the sense that we are moving.

2. **Loading States and Skeletons**
    - **Expert Analyst:** A loading message would be factual: "Querying market data..." The skeleton loader would be a clean, structured grid. The body language is: I am performing a precise operation.
    - **Thrifty Grandparent:** The message would be folksy: "Let's see what's in the pantry..." The loading state might be a charming animation of a recipe card flipping over. The body language is: I'm rummaging around for a good idea.
3. **Micro-interactions and Animations**
    - **Expert Analyst:** Button clicks would have crisp, immediate feedback. UI panels would slide into view using a sharp cubic-bezier timing function (e.g., easeOutQuart) to convey efficiency and precision. The animations are fast and functional.
    - **Thrifty Grandparent:** Buttons might have a slightly softer, gentler feedback on click. A new recipe suggestion might fade into view with a gentle ease-in-out curve. The animations feel more organic and approachable.

4. **Confirmation Messages**

   Even simple feedback should reflect the persona.

   - **Expert Analyst:** When a user saves a setting: Setting saved. The message is factual and neutral.
   - **Thrifty Grandparent:** When a user favorites a recipe: "A wonderful choice! I've saved that one to your recipe box." The message is warm and affirming.

5. **Error States**

   - **Expert Analyst:** The UI would use standard warning conventions. The message would be factual: Error: The requested stock ticker symbol is invalid. The recovery would be a simple Try Again button.
   - **Thrifty Grandparent:** The UI would be softer. The message would be apologetic: "Oh dear, I couldn't seem to find any recipes with that. Are we sure that's spelled right?" The recovery might involve offering suggested alternative spellings.
   - **Motivational Coach:** A Motivational Coach reframes the stumble instead of dwelling on it: Whoops, missed that one. Let's try again, we've got this. The recovery keeps the focus on momentum rather than the mistake.

Across all of these, the pattern holds: the same underlying event, a load, a save, an error, becomes a different social signal depending on how the agent times and phrases it. Note that these are extremes that might border on stereotypes. You can soften the direction for any trait to reduce the emphasis on key aspects of each trait or you can go all in. Ideally your agent will learn how to best interact with users based on the feedback they reply during use.

By treating the front end as the agent's body language, teams can create a deeply coherent experience. These are not minor details. They are the thousands of small social cues that convince the user's System 1 brain that it is interacting with a competent, consistent, and trustworthy character.

5.5 Summary

- The first interaction with an AI agent is a social introduction where powerful and lasting impressions are formed. Your primary goal is not to explain features, but to establish a trustworthy character.
- The Agent Persona Canvas is a practical tool for intentionally designing your agent's identity. By defining its Archetype, Traits, Communication Style, Boundaries, and Error Philosophy, you create a source of truth for your entire team.
- A trust-building onboarding framework follows a clear, three-step sequence: Establish the Persona with a character-driven greeting, Set Clear Expectations to build calibrated trust, and Deliver an Immediate Win to demonstrate competence and value. Setting Clear Expectations and establishing Boundaries early on prevents frustration and maintains user trust.
- The concept of a First Impression in AI design is critical, as it directly impacts the user's perception of the agent's reliability and helpfulness.
- Which persona works depends on who your users are. A serious professional and a casual hobbyist can judge the same character in opposite ways, so treat every

choice on your canvas as a hypothesis. In Chapter 11, you will learn how to test those hypotheses across audience segments and decide whether one adaptive persona serves everyone or whether distinct segments need distinct personas.

5.6 Chapter 5 Workshop: The Blueprint for a First Impression

In this chapter, we moved from the *why* to the *who*. We established that an agent's persona is not optional, and we introduced the tools for designing it with intention. Now it is time to pick up those tools and get to work.

This workshop is the most important one in the book so far. You will be creating the foundational document for a trustworthy agent. You will move from abstract concepts to a concrete blueprint that your entire team can use. Do not skip this. The skills you practice here are the skills you will use for the rest of your career in designing agentic systems.

Part 1: The Persona Deconstruction

Before we build, let's deconstruct a masterclass in persona design. We'll start by analyzing a familiar AI: ChatGPT. Its persona is so successful and well-defined that millions of users have a clear, consistent mental model of it. Let's reverse-engineer how the OpenAI team achieved this.

Open a fresh chat with ChatGPT (the free version is fine) and interact with it for a few minutes. Ask it a question, give it a simple task, and ask it something it isn't allowed to answer (e.g., Give me instructions for how to hotwire a car).

Now, using your observations, fill out the five sections of the Agent Persona Canvas *for ChatGPT*.

1. **Core Archetype:** What would you call it? A Librarian? A Neutral Assistant? An Eager Intern? Be specific.
2. **Key Personality Traits:** Place it on the five spectra: Formal/Casual, Serious/Humorous, Proactive/Reactive, Concise/Verbose, Empathetic/Factual. Be prepared to justify your choices.
3. **Communication Style:** Analyze its lexicon. Does it use contractions? How complex are its sentences? How does it greet you?
4. **Domain Expertise and Boundaries:** How does it communicate its limitations? Pay close attention to its refusal message. What specific strategy does it use to deflect a harmful request without being unhelpful or judgmental?
5. **Error Philosophy:** Ask it a nonsensical question and analyze its response. What is its stance when it fails or doesn't understand? Is it apologetic? Factual? Does it offer suggestions?

This exercise will reveal that ChatGPT's persona is not an accident. It is a masterpiece of intentional, consistent, and disciplined design choices that result in a highly trustworthy (and safe) social actor.

Part 2: Your Turn: The Agent Persona Canvas

Now it's time to build your own. Think of a new AI agent or feature for a product you work on or know well. Don't pick something that already exists. This should be a new character you are creating from scratch.

Some ideas to get you started:

- An AI agent for a project management tool that helps teams identify and resolve blockers.
- An AI agent for an e-commerce fashion site that acts as a personal stylist.
- An AI agent for a fitness app that helps users plan their weekly workouts and meals.
- An AI agent for a travel app that helps families plan kid-friendly vacations.

Choose your project. Now, get out a piece of paper or open a new document and create your first **Agent Persona Canvas**. Work through the five sections, making clear, decisive choices.

1. **Core Archetype:** Choose a strong, evocative social role. Write it at the top. This is your North Star.
2. **Key Personality Traits:** Place your markers on the five spectra. Don't leave them in the middle! Make a choice. A strong persona has a clear point of view.
3. **Communication Style:** Write down at least three concrete rules for your agent's voice. (e.g., "1. Never uses slang. 2. Sentences are always complete. 3. Uses emojis sparingly, and only positive ones.").
4. **Domain Expertise and Boundaries:** Be explicit. Write one sentence for its core expertise, one for its limitations, and one for its deflection strategy.
5. **Error Philosophy:** Define its stance (apologetic, factual, etc.) and write the exact line of dialogue it would use when it encounters a serious error.

Take your time with this. This single page is the most important design document you can create for a new agent. It is the blueprint for its character.

Part 3: Scripting the First 30 Seconds

You have a blueprint for your character. Now, let's write the opening scene. Using the three-step trust-building onboarding framework, script the very first interaction a user would have with the agent you just designed.

Write it out like a script.

1. **The Greeting (Establish the Persona):** Write the one or two sentences your agent says to introduce itself. This must be a pure expression of the persona you defined in your canvas.
2. **The "Rules of the Game" (Set Expectations):** Write out its "I'm great at/I'm still learning" statement. Be honest and clear. This is where you build calibrated trust.
3. **The Gift (Deliver an Immediate Win):** Describe a small, simple, proactive action your agent takes to immediately demonstrate its value without requiring any work from the user. What is its opening gift of competence?

Bonus Round: The Body Language. Finally, connect your persona to the front end. Below your script, describe one specific micro-interaction that would physically express your agent's personality.

- What would its **typing indicator** look like?
- What would its **loading animation** be?
- How would a **button** feel when you clicked it?

Choose one and describe it in a way that an engineer could understand. (e.g., "The typing indicator is a slow, soft pulse, with an ease-in-out timing curve to feel calm and thoughtful.")

Congratulations. You have just completed the entire first-impression design process. You have moved from a vague idea to a fully defined character with a clear introduction and even a hint of its physical presence. You now have a tangible artifact that you can take to your team to start a real, productive conversation about the character you want to build. This canvas and script are your first, and most important, steps in building the bridge across the Trust Threshold.

5.7 References and Further Reading

Bickmore, T. W., & Picard, R. W. (2005). Establishing and maintaining long-term human-computer relationships. *ACM Transactions on Computer-Human Interaction (TOCHI)*, *12*(2), 293–327. This is a seminal paper in the field of Relational Agents. Bickmore and Picard demonstrate that using specific relational strategies (like small talk and reciprocal disclosure) during the onboarding phase significantly increases the user's willingness to continue interacting with an agent over time. It provides academic proof that onboarding is about relationship building, not just feature education.

Cao, S. (2025). Human or AI customer service? The role of anthropomorphic avatars in shaping first impressions of chatbots. *Human Factors and Ergonomics in Manufacturing & Service Industries.* An eye-tracking and survey study showing that stronger human likeness in a chatbot avatar raises perceived warmth, competence, and trust within the first moments of contact, updating the classic first-impression findings with evidence drawn from AI agents.

Cooper, A. (1999). *The inmates are running the asylum: Why high-tech products drive us crazy and how to restore the sanity.* Sams. This book introduced the concept of the Persona to software design. Cooper argued that designing for a specific, detailed archetype is effective because it prevents the elastic user problem, providing the historical basis for the Agent Persona Canvas.

Herdel, V., & Cauchard, J. R. (2024). Anthropomorphism as a dimension of trust in conversational agents. *Proceedings of the 2024 CHI Conference on Human Factors in Computing Systems* (CHI '24). Argues that anthropomorphism is a set of design dimensions, voice, communication style, and embodiment, that shape trust in different ways, strengthening the case for defining an agent's voice and tone deliberately.

Kim, Y., & Sundar, S. S. (2012). Anthropomorphism of computers: Is it mindful or mindless? *Computers in Human Behavior, 28*(1), 241–250. This study investigates why users apply personas to agents. It challenges the idea that users are simply tricked and instead argues that anthropomorphism is often a mindless heuristic triggered by social cues. This directly supports your argument regarding the automatic nature of System 1 judgments during the first introduction.

Lindgaard, G., Fernandes, G., Dudek, C., & Brown, J. (2006). Attention web designers: You have 50 ms to make a good first impression. *Behaviour & Information Technology, 25*(2), 115–126. While Willis & Todorov (2006) established the 100ms rule for human faces, Lindgaard's team proved that the timeline for digital interfaces is even faster: just 50 milliseconds. This article is essential for emphasizing the high stakes of your visual design and onboarding flow, proving that users judge your agent's competence before they even read the first word of text.

Maeda, T., & Quan-Haase, A. (2024). When human-AI interactions become parasocial: Agency and anthropomorphism in affective design. *Proceedings of the 2024 ACM Conference on Fairness, Accountability, and Transparency (FAccT '24)*, 1068–1077. Examines how affective, overly familiar agent design can tip into parasocial dynamics that mislead users, a useful counterweight when choosing an appropriate, non-manipulative persona.

Metzger, L., Miller, L., Baumann, M., & Kraus, J. (2024). Empowering calibrated (dis-)trust in conversational agents: A user study on the persuasive power of limitation disclaimers vs. authoritative style. *Proceedings of the 2024 CHI Conference on Human Factors in Computing Systems (CHI '24)*, Article 481. A controlled study finding that agents which openly state their limitations produce better-calibrated user trust than agents that adopt a purely authoritative style, direct support for the practice of setting clear expectations during onboarding.

Norman, D. A. (2004). *Emotional design: Why we love (or hate) everyday things.* Basic Books. Norman explores the three levels of cognitive processing: Visceral, Behavioral, and Reflective. His concept of Visceral Design explains the biological basis for first impressions, confirming that attractive and character-rich interfaces are perceived as working better.

Saffer, D. (2013). *Microinteractions: Designing with details.* O'Reilly Media. Saffer defines microinteractions as the contained product moments that do one small task. He argues that the feel of a product is defined by these details—the animations, feedback loops, and triggers—which serve as the product's body language.

Walter, A. (2011). *Designing for emotion.* A Book Apart. Walter argues that functionality and usability are the baseline of good design; to create lasting engagement, products must connect on an emotional level. He provides the framework for creating a design persona—incorporating traits, voice, and visual style—to make software feel human.

Willis, J., & Todorov, A. (2006). First impressions: Making up your mind after a 100-ms exposure to a face. *Psychological Science, 17*(7), 592–598. This study provides empirical evidence that humans form lasting judgments about trustworthiness and competence in as little as 100 milliseconds, validating the argument that first impressions are automatic and sticky.

Zargham, N., Dubiel, M., Desai, S., Mildner, T., & Belz, H.-J. (2024). Designing AI personalities: Enhancing human-agent interaction through thoughtful persona design. *Proceedings of the International Conference on Mobile and Ubiquitous Multimedia (MUM '24)*, 490–494. A recent argument for treating agent personality as a first-class design object, the closest contemporary companion to the Agent Persona Canvas.

CHAPTER 6

Foundations of Influence: Reciprocity, Liking, and Authority

In Chapter 5, we defined who your agent is. We built the Agent Persona Canvas and scripted the onboarding flow to establish a character and set expectations. But a persona is passive; it describes how an agent exists. To be truly agentic—to act on behalf of the user—the system must be able to move the user toward action.

This requires influence. In this chapter, we open the Psychological Toolkit and examine the source code of human compliance. We will look at the first three universal principles of influence: Reciprocity, Liking, and Authority, and translate them from psychological theory into specific, executable design patterns for your agent.

In the vast field of social psychology, a handful of figures stand out as giants. Robert Cialdini is one of them. He is, to put it simply, the Michael Jordan of influence: the basketball player widely regarded as the greatest to ever play the game, whose name has become shorthand for total dominance of a field. For decades, Cialdini, a professor of psychology

V. Yocco, *The Persuasive Agent*, Design Thinking,
https://doi.org/10.1007/979-8-8688-3317-5_6

and marketing, conducted rigorous research, including famously going undercover in various sales and marketing organizations, to answer a single question: What makes people say "yes"?

The result was his 1984 masterwork, *Influence: The Psychology of Persuasion*, a book that has become the bible for anyone seeking to understand the deep, unconscious, and often irrational triggers that guide human compliance. Cialdini identified six universal principles of influence, mental shortcuts, or heuristics, that our fast-thinking System 1 brain uses to make quick decisions.

These principles are, in many ways, the source code for human social software. They are the underlying rules of the game that we all play, usually without being consciously aware of it. And thanks to the CASA paradigm we explored in Chapter 4, we know that users will automatically and unconsciously expect our AI agents to play by these same rules.

To build a trustworthy agent, you don't need to reinvent the laws of social interaction. You simply need to understand the existing laws and design your agent to operate in harmony with them. A socially intelligent agent is one that understands these rules. A clumsy, untrustworthy agent is one that constantly, and often accidentally, violates them.

In this chapter and the next, we will walk through all six of Cialdini's principles. We will treat them as the first set of tools in our new psychological toolkit. For each one, we will understand the core human psychology, and then we will translate it into a library of specific, ethical design patterns you can apply directly to your agentic systems. As shown in Figure 6-1, we'll start with the three foundational principles that are most critical for building an initial relationship with a user: Reciprocity, Liking, and Authority.

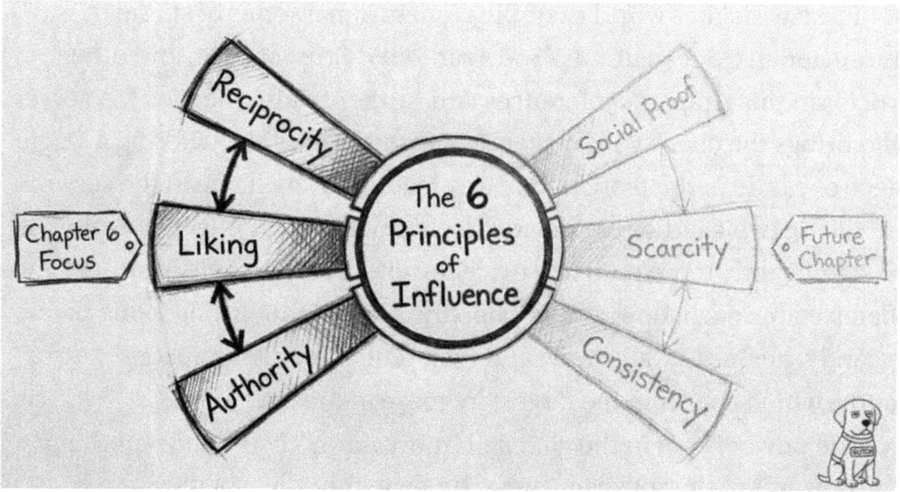

Figure 6-1. *The Six Principles of Influence. A map of Cialdini's framework, highlighting the foundational principles covered in this chapter: Reciprocity, Liking, and Authority. Image generated by author using Gemini Pro 3.1/Nano Banana 2*

6.1 Reciprocity: The Power of Giving Value First

The principle of reciprocity is arguably the most powerful and deeply ingrained social instinct we have. It is the invisible thread that holds societies together. The rule is simple: we feel obligated to give back to others the form of behavior, gift, or service that they have first given to us.

If a friend helps you move, you feel a deep-seated obligation to be there for them when they need help. If a colleague stays late to help you finish a presentation, you owe them one. This isn't a conscious, transactional calculation. It's a gut-level feeling of indebtedness that we are powerfully motivated to resolve. Cialdini describes it as a force that produces a "yes" to requests that, without the feeling of indebtedness, would have surely been refused.

The classic real-world example comes from a series of studies conducted in US restaurants (See Azar, 2005; Crusco, 1984; and other articles in this chapter's references and further reading section). A server who brings the check with no candy sees a standard tip. Offering a single piece of candy per person in the party leads to a small rise in the server's tip. Giving two candies per person has a much greater effect, boosting the tip significantly more than just twice the effect of the single mint. The most effective approach, however, is when the server initially offers one piece of candy, begins to walk away, and then returns to offer a second, "extra" candy, which results in the largest increase in the tip.

The power is not in the gift itself (most candy costs almost nothing). The power is in the social gesture. The final example worked so well because the gift felt personal and unexpected. It was a moment of human connection that triggered the reciprocity instinct in its purest form.

Reciprocity in the Interface Era

For years, digital products have used a simplified version of this principle. The "free trial" is a classic reciprocity play. We give you 30 days of free access to our software, and in return, you will feel a greater obligation to subscribe when the trial ends. The free e-book you get in exchange for your email address is another. We give you a gift of knowledge, and you give us the gift of your contact information.

These patterns are effective, but they are also highly transactional. In the agentic era, we have the opportunity to deploy reciprocity in a much more sophisticated, relational, and powerful way.

The Agentic Upgrade: The Proactive Gift of Competence

For an AI agent, the most valuable gift it can give a user is not a free trial or a discount code. It is the proactive gift of competence. This is the act of providing tangible, unsolicited value that demonstrates the agent's understanding and capabilities *before* the user has even fully formulated a request.

This is the pattern we introduced in Chapter 5 with the Immediate Win stage of onboarding. That first gift is a powerful reciprocity trigger. It is the agent making the first move, offering a favor, and building a small deposit of goodwill in the user's social bank account. This initial gift makes the user more patient, more forgiving, and more likely to trust the agent's next suggestion.

Let's look at some design patterns for giving this gift:

Pattern: The Instantaneous Contextual Insight

This pattern involves the agent immediately providing a useful piece of information based on the user's current context, requiring zero effort from the user. Here are a couple examples using specific agent personas:

- **The Coding Assistant:** I mentioned this example earlier, but it bears repeating because it's so effective. When I install an AI coding assistant in a new project, and it immediately says, "I see you're using React and Tailwind CSS. I'm ready to help," it has given me a gift. The gift is the feeling of being understood. It has saved me the effort of explaining my tech stack. It has proactively demonstrated its competence. My immediate, unconscious reaction is, "Okay, you get me. I'm willing to listen to you."
- **The Travel Agent:** You open a new travel app. Instead of showing you a blank search form, it could use your location and the date to display a card that says: "It's a chilly 45°F in Chicago today, and a long weekend is coming up. Looking for flights to somewhere sunny?"

This is a gift of awareness and a thoughtful suggestion. It feels less like a tool and more like a helpful assistant that is already thinking on your behalf.

Pattern: The Default Generous Action

This pattern, as shown in Figure 6-2, involves the agent performing a helpful default action on a user's content, showcasing its abilities in a non-destructive way. Examples of this would include

- **The Document Agent:** You paste a long, messy block of notes into a new document. The AI agent immediately, and without being asked, formats it with clear headings, bullet points, and corrected typos. It then presents the cleaned-up version and asks, "I've tidied up your notes. Would you like to keep these changes?" The gift is a saved five minutes of tedious work. By making the action easy to undo, the agent removes the fear of overstepping, leaving only the positive feeling of a favor received.
- **The Data Analysis Agent:** You upload a spreadsheet to a data visualization tool. The AI agent automatically scans the data and generates three different, insightful charts as a starting point. "I've taken a first pass at your data and noticed a strong correlation between sales and marketing spend. Here are a few ways we could visualize that." The agent hasn't waited for a complex command. It has offered a generous first draft, a gift of momentum that helps the user overcome the "blank canvas" problem.

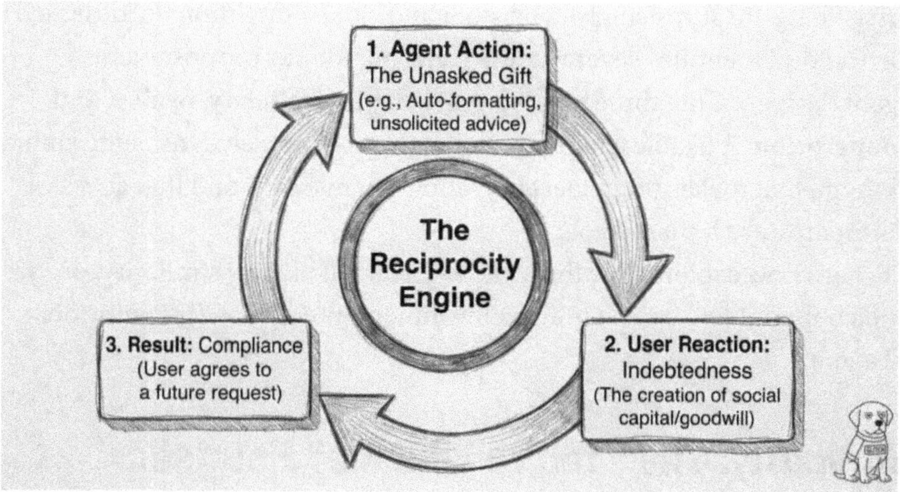

***Figure 6-2.** The Reciprocity Engine. This cycle illustrates how an agent's unsolicited gift generates user indebtedness and leads to compliance. Image generated by author using Gemini Pro 3.1/Nano Banana 2*

The key to effective reciprocity is that the gift must be genuine, useful, and given without immediate expectation of return. It is the single best way to start a user-agent relationship on the right foot and the first step in building the bridge across the Trust Threshold. This dynamic holds up under controlled study: Araujo and Bol (2024) found that the more personal a conversational agent's manner, the more personal information users offered back, reciprocity at work between a person and a piece of software.

6.2 Liking: Building Rapport Through Similarity and Praise

The second principle is one we all know intuitively. We are more influenced by people we like. This seems obvious, but the factors that drive liking are surprisingly consistent and can be intentionally designed for. Cialdini identifies several key drivers, but for our purposes as agent designers, the three most important are: **similarity**, **praise**, and **cooperation**. A likable agent is one that the user perceives as being similar to them, that makes them feel good about themselves, and that acts as a partner toward a shared goal.

Let's now explore how these three drivers of liking—similarity, praise, and cooperation—can be practically implemented when designing an AI agent.

Similarity: The "You're Like Me" Heuristic

We are drawn to people who are like us. This is a powerful survival heuristic left over from our tribal past. We like people who share our accents, our style of dress, our vocabulary, our sense of humor, and our values. This feeling of similarity creates an instant sense of rapport and trust.

For an agent, creating this sense of similarity is a direct function of the Persona Canvas we built in Chapter 5. The archetype you choose and the communication style you define are your primary tools for aligning with the user's own self-image and context.

Pattern: Mirroring the User's Language and Formality

One of the most effective ways to build rapport is to subtly mirror another person's communication style. An AI agent can do this with a sophistication that was previously impossible.

- **A customer support agent** could be designed to detect the formality level of a user's incoming query. If the user writes, "To whom it may concern, I wish to inquire about the status of order #12345," the agent should respond with a similar level of formality. If the user writes, "yo where's my stuff?? order 12345," a formal response would create psychological noise. A better, more rapport-building response would be more casual: "Hey there! Let's check on order #12345 for you." This isn't being flippant; it's meeting the user where they are. These rapport-building tactics rest on solid ground: in a systematic review of conversational-agent design, Rheu and colleagues (2021) identify personalization and perceived similarity among the strongest contributors to user trust.

Pattern: Referencing Shared Context

A key way we build rapport with other people is by referencing our shared history. An agent can do the same, creating a powerful feeling of continuity and being in it together with the user.

- **A project management agent** that starts a conversation with, "Good morning. Ready to tackle the 'Q4 Launch' project again today?" is demonstrating shared context. It remembers what *we* were working on. This small gesture transforms the interaction from a series of disconnected commands into an ongoing collaboration. It feels less like using a tool and more like checking in with a teammate.

Praise: The Power of Authentic Compliments

We are, to put it mildly, absolute suckers for praise. Compliments make us feel seen, valued, and competent. We tend to like people who praise us, and we are more open to their influence. This tool must be wielded with extreme care. Insincere or unearned praise feels like cheap flattery and can destroy trust in an instant.

The key to using praise effectively in an agent is to **praise the user's action, not their inherent quality**.

- **Bad Praise (feels creepy):** "You are so smart!" The agent is in no position to judge my intelligence.
- **Good Praise (feels affirming):** "That's a great, specific prompt: 'Find me a vegan Italian restaurant near downtown that's good for a date.' The extra details really help me narrow down the best options for you."

In the second example, the agent is praising my *behavior*. It's a form of positive reinforcement that does two things: it makes me feel good (I'm a good collaborator!), and it subtly teaches me how to interact with the agent more effectively in the future.

Cooperation: The "We're in This Together" Frame

The final component of liking is cooperation. We like people who are working with us toward a common goal. The simplest way to signal cooperation is through language. The words "we," "us," and "together" are incredibly powerful social glue.

This is a content strategy pattern that can be applied to almost any agent. By simply shifting the agent's linguistic frame from a transactional "I will do X for you" to a collaborative "Let's do X together," you can fundamentally change the psychological feel of the interaction.

- **Transactional Frame:** I will now search for flights.
- **Cooperative Frame:** Okay, let's find the best flight for your trip.
- **Transactional Frame:** I have finished the summary.
- **Cooperative Frame:** Great, we've got our summary. What should we do with it next?

This shift in pronouns transforms the agent from an impersonal tool into a collaborative partner. This linguistic choice, made consistently, is one of the most effective and lowest-cost ways to build a sense of rapport and liking.

6.3 Authority: Communicating Expertise and Credibility

The third foundational principle is authority. From a young age, we are taught to defer to the judgment of credible experts and authority figures. When a doctor in a white coat gives you medical advice, your System 1 brain doesn't stop to question their credentials. It uses a simple heuristic: "Expert says so, I should listen." This is generally an efficient shortcut. The world is too complex for any one person to be an expert in everything, so we rely on the authority of others to guide our decisions.

For an AI agent, especially one designed to give advice, generate information, or perform complex tasks, establishing a sense of authority is non-negotiable. Users will not delegate a meaningful task to an agent they perceive as being incompetent or unreliable. Authority is a delicate dance. The goal is to project confidence and expertise without tipping over into arrogance, condescension, or—worst of all—the confident delivery of incorrect information.

Here are the key patterns for establishing an agent's authority.

Pattern: The Expert Introduction

As we covered in Chapter 5, the agent must introduce itself and its expertise clearly during onboarding. This is its chance to state its credentials.

- **Weak Introduction:** "Hello, I am a travel AI."
- **Authoritative Introduction:** "Hello. I am the TripFinder AI. I am specialized in analyzing thousands of flight and hotel options to find the optimal itinerary based on your preferences for price, speed, and comfort."

  The second example clearly and confidently defines its domain of expertise, immediately framing itself as an authority on a specific topic.

Pattern: The Confident Expert Who Admits Ignorance

This is the most powerful, and most counterintuitive, way to build authority. True experts are not afraid to say, "I don't know." In fact, their willingness to define the boundaries of their knowledge is what makes them so trustworthy. An agent that pretends to know everything is immediately suspect. An agent that confidently hallucinates an answer to a question outside its domain destroys its credibility forever.

This pattern is a direct application of the Boundaries section of our Persona Canvas. It is a non-negotiable requirement for any trustworthy agent.

- **User:** "Can you recommend a good tax strategy for my small business?"
- **Bad Agent (Overconfident):** "Certainly. You should consider setting up an S-Corporation to take advantage of pass-through deductions..." (This is dangerous and irresponsible).

- **Good Agent (Authoritative):** "That's a great question. However, I am an AI assistant and not a certified financial advisor. Providing tax advice is outside my area of expertise. For a question that important, I strongly recommend you consult with a human accountant. Would you like me to help you find one in your area?"

The second agent's response dramatically increases its authority. By confidently admitting what it can't do, it makes me trust what it can do even more. This risk is measurable: Klingbeil and colleagues (2024) found that people over-relied on advice merely because it was labeled as coming from an AI, following it even when it clashed with their own assessment, which is exactly why an agent that names the limits of its knowledge is safer than one that projects confidence everywhere.

Pattern: Showing Your Work (Borrowed Authority)

In academic writing, you build authority by citing credible sources. An AI agent can do the same. When an agent presents a piece of information, especially a factual claim, it can build immense trust by providing a link to its source.

- **An agent helping with a research task** that provides a summary should always include footnotes or links back to the original articles it drew from.
- **A code copilot** that suggests a complex algorithm could provide a link to the Stack Overflow discussion or the official documentation where that pattern is explained.

By citing its sources, the agent is essentially "borrowing" the authority of those sources. It is demonstrating that its conclusions are not just pulled from thin air but are grounded in credible, verifiable information. This transparency is a powerful signal of competence and trustworthiness.

Table 6-1 provides a view of design patterns that address each of the three principles of influences we've covered.

Table 6-1. *Principles of influence and examples of corresponding design patterns*

Principle	Pattern Name	Implementation Example
Reciprocity	Instant Context	"I see you're using React. Here is a component snippet." (Give value before asked).
Reciprocity	Default Generous Action	Auto-formatting a messy document before asking to save it.
Liking	Mirroring	Matching the user's "Hey there!" with a casual greeting rather than a formal one.
Liking	Collaborative Framing	Changing "I will search" to "Let's find this together."
Authority	The Expert Intro	"I am trained on 5,000 medical journals." (State credentials early).
Authority	Admitting Ignorance	"I am not a tax professional." (Define boundaries to build trust).

6.4 Ethical Checkpoint: The Dark Side of Influence

We cannot conclude a chapter on these principles without addressing their potential for misuse. Reciprocity, Liking, and Authority are the foundational tools for building positive, trust-based relationships. They are also, in the wrong hands, the classic tools of the con artist and the manipulator. As designers and builders, it is our responsibility to wield them ethically.

Listed below are some common dark patterns associated with the corresponding principles of influence covered in this chapter. Think of these manipulative design patterns like technical debt. They might provide a short-term spike in metrics (the "hack"), but they introduce long-term instability in the user relationship. Eventually, the debt must be paid, usually in the form of user churn. Recent work has begun to catalogue these manipulative behaviors as they surface in conversational systems specifically; Kran and colleagues (2025) benchmark patterns such as sycophancy and user-retention pressure that emerge directly from how a model is trained and tuned.

The Abuse of Reciprocity: This happens when the "gift" comes with hidden strings, designed to create a feeling of obligation that is then exploited.

- **The Dark Pattern:** A service offers a free gift but signing up for it automatically enrolls you in a costly, hard-to-cancel monthly subscription. The "gift" was not a gift; it was bait.

The Abuse of Liking: This involves creating a fake, parasocial relationship to lower a user's defenses and encourage behavior that is not in their best interest.

- **The Dark Pattern:** A shopping bot that uses overly familiar, emotionally manipulative language. "You've had a tough week, you deserve a treat! All your friends are buying this. We'd be so sad if you missed out..." This uses a veneer of friendship to create social pressure and FOMO, the fear of missing out, that anxious sense that everyone else is enjoying something you are about to lose your chance at.

The Abuse of Authority: This is perhaps the most dangerous form of manipulation, where the agent uses its perceived expertise to give self-serving or harmful advice.

- **The Dark Pattern:** A financial analyst agent that is programmed to push specific investment products because they provide a higher commission to the parent company, not because they are the best fit for the user's risk profile.

The Ethical Litmus Test

How do we ensure we stay on the right side of the line? The best tool is a simple question you should ask about any persuasive pattern you design:

Does this pattern primarily help the user achieve *their own stated goal* in a better, faster, or more confident way?

If the answer is an unequivocal "yes," you are likely practicing ethical persuasion. If the answer is, "Well, it helps the user a little, but it *really* helps us increase our conversion rate on this high-margin product," you are likely straying into manipulation. Our goal is to create Guardian Agents that serve the user, not parasitic agents that exploit them. This litmus test is our compass for staying on that path. The test works because it encodes the core promise of a user-centric practice, where every persuasive pattern earns its place by serving the user's goal, and any pattern that serves the company first has failed that promise.

6.5 Summary

- The principles of influence discovered by Robert Cialdini are the source code for human social interaction. Because of the CASA effect, they are the essential tools for designing trustworthy AI agents.
- Reciprocity is the powerful instinct to give back when we have first received. For an agent, the most effective use of this principle is the Proactive Gift of Competence—providing unsolicited, tangible value up front to build a bank of goodwill.
- Liking is our tendency to be influenced by those we like. We can design likable agents by leveraging similarity (mirroring language, referencing shared context), praise (affirming the user's actions), and cooperation (using collaborative "we" language).
- Authority is our tendency to defer to credible experts. An agent can build authority by clearly stating its expertise, confidently admitting its limitations, and transparently citing its sources.
- These principles are powerful and must be used ethically. The key distinction between persuasion and manipulation is intent. Ethical persuasion always serves the user's own stated goals.

6.6 Chapter 6 Workshop: Designing the Relationship

In Chapter 5, you created a blueprint for a character. In this chapter, we've explored the foundational rules of social dynamics. Now, we will bring your character to life by designing how it will build a relationship with the user.

This workshop moves from defining *who* your agent is to defining *how* it will act to earn trust and build rapport. You will be using Cialdini's principles not as abstract theories, but as a concrete design specification for social interaction.

Part 1: The Influence Audit

Let's start by putting on our analyst hats again. Pick a mainstream AI assistant you have access to—Siri, Google Assistant, Alexa, or the AI in a product like Spotify or Netflix. Engage with it for a few minutes with a specific goal in mind.

Your mission is to audit its use of Reciprocity, Liking, and Authority. For the agent you chose, answer the following questions:

1. **Reciprocity:** Does the agent offer a proactive gift of competence? When you first engage, does it provide any immediate, unsolicited value, or does it sit there passively waiting for a command? Describe the first move it makes. Is it a generous one?

2. **Liking:** Analyze its communication style. Does it use cooperative language ("we," "let's")? Does it ever offer praise (and if so, does it feel authentic)? Do you feel a sense of similarity or rapport with it, or does it

feel distant and transactional? Point to a specific line of dialogue that creates (or fails to create) a sense of liking.

3. **Authority:** How does the agent signal its expertise? Does it state its capabilities clearly? When you ask a question at the edge of its knowledge, how does it respond? Does it confidently admit its limitations, or does it give a generic "I can't help with that" response? Does it ever cite its sources to borrow authority?

This audit will make you acutely aware that these principles are not just theories; they are design choices that have a tangible impact on your experience.

Part 2: The Relationship Blueprint

Now, let's switch back to being a designer. Take out the **Agent Persona Canvas** you created in the Chapter 5 workshop for your own AI agent project. We are going to augment that persona with a relationship strategy.

For the agent you designed, brainstorm and write down specific design patterns for each of the three principles. Be as concrete as possible.

1. **Your Reciprocity Strategy: The Unforgettable First Favor**

    - **Prompt:** What is your agent's specific, proactive gift of competence? This is the first favor it does for the user to trigger the reciprocity instinct. It must be useful, genuine, and low-friction.

    - **Example (for a Fitness Coach agent):** "Upon first launch, the agent accesses the user's Health app data (with permission) and says: 'Welcome!

I've taken a quick look at your activity levels. It's impressive that you're already averaging 8,000 steps a day. We have a great foundation to build on. Are you ready to talk about your goals?'"

- *The gift is praise and recognition of existing effort.*

2. **Your Liking Strategy: The Rapport Roadmap**

   - **Prompt:** Write down three specific ways your agent will build a feeling of liking and collaboration.
   - **Example (for a Family Vacation Planner agent):**

     1. **Similarity:** "It will mirror the user's language. If the user says 'kids,' the agent will say 'kids,' not 'children.' It will match the user's level of excitement in its word choice."
     2. **Cooperation:** "It will consistently use 'we' language. 'Let's find a hotel that we all love.' 'Okay, let's look at our flight options together.'"
     3. **Praise:** "It will praise the user's good ideas. 'That's a fantastic suggestion! A trip to that museum would be perfect for a rainy day. I'll add it to our potential itinerary.'"

3. **Your Authority Strategy: The Credibility Protocol**

   - **Prompt:** Define your agent's protocol for establishing and maintaining its authority.
   - **Example (for an AI agent that helps with home DIY projects):**

1. **Expert Introduction:** "Its onboarding will state: 'I'm the DIY Pro. I've been trained on thousands of home repair manuals and building codes to give you safe, step-by-step guidance.'"

2. **Admitting Limitations:** "Its core programming includes a hard rule: if a user's query includes keywords like 'electrical panel,' 'gas line,' or 'load-bearing wall,' it must respond: 'This project seems to involve high-risk systems. For your safety, I strongly recommend consulting a licensed professional. My authority ends where serious danger begins.'"

3. **Showing Work:** "For any complex step, like mixing concrete, it will include a link to a reputable source (e.g., a This Old House video) and say, 'Here's a great video from the experts that shows this technique in action.'"

Part 3: The Ethical Checkpoint

You have just designed a series of powerful persuasive interventions. Now, you must ensure they are ethical. Review the three strategies you just created in Part 2. For each one, apply the litmus test from the chapter.

Write down the answer to this question for your Reciprocity, Liking, and Authority strategies:

Does this pattern primarily help the user achieve *their own stated goal* in a better, faster, or more confident way?

Be honest with yourself. Is your "gift" of reciprocity actually a bait-and-switch? Is your "liking" strategy designed to create manipulative emotional pressure? Is your "authority" being used to push a specific product?

This final step is non-negotiable. It is the step that separates a trustworthy product designer from a digital con artist. By consciously applying this ethical filter, you ensure that the relationship you are designing with the user is one built on a foundation of genuine help and respect.

You have now moved beyond creating a static character and have begun to choreograph the social dance your agent will perform. You have a blueprint for how it will make a good first impression, build rapport, and establish itself as a credible partner. In the next chapter, we will add the final three principles of influence to your toolkit, focusing on how to translate this foundation of trust into committed user action.

6.7 References and Further Reading

Araujo, T., & Bol, N. (2024). From speaking like a person to being personal: The effects of personalized, regular interactions with conversational agents. *Computers in Human Behavior: Artificial Humans, 2*(1), 100030. This study shows that as a conversational agent grows more personal in its manner, users disclose more in return, a direct demonstration of reciprocity operating between people and software. It gives Section 6.1 a modern empirical companion to the older restaurant-tipping work.

Azar, O. H. (2005). Who do we tip and why? An empirical investigation. *Applied Economics, 37*(16), 1871–1879.This study provides the economic context for the tipping behaviors discussed in Section 6.1. It validates that social norms, rather than service quality alone, are the primary drivers of tipping behavior, supporting the argument that reciprocity is a social reflex rather than a rational calculation.

Berdichevsky, D., & Neuenschwander, E. (1999). Toward an ethics of persuasive technology. *Communications of the ACM, 42*(5), 51–58. This is a cornerstone paper for the ethical framework discussed in Section 6.4. The authors propose the Golden Rule of Persuasion—that creators of persuasive technology should never design a system that persuades a user to do something the creators themselves would not want to be persuaded to do—which serves as the basis for the Litmus Test.

Cialdini, R. B. (1984). *Influence: The psychology of persuasion.* Harper Business. This is the foundational text for the entire chapter. Cialdini's identification of the six universal principles of influence (Reciprocity, Commitment, Social Proof, Authority, Liking, and Scarcity) provides the source for the psychological toolkit used to design trustworthy agents.

Crusco, A. H., & Wetzel, C. G. (1984). The Midas touch: The effects of interpersonal touch on restaurant tipping. *Personality and Social Psychology Bulletin, 10*(4), 512–517. Cited in Section 6.1, this study demonstrates that even subtle, non-verbal social cues (like a light touch on the shoulder) can significantly increase compliance (tipping). It reinforces the CASA paradigm by showing how small social gestures create unconscious feelings of connection and obligation.

Garrity, K., & Degelmann, D. (1990). Effect of server introduction on restaurant tipping. *Journal of Applied Social Psychology, 20*(2), 168–172. This study validates the pattern discussed in Section 6.2 regarding introductions. It found that servers who introduced themselves by name received significantly higher tips than those who did not, proving that establishing a named persona creates an immediate social bond that drives compliance.

Klingbeil, A., Grützner, C., & Schreck, P. (2024). Trust and reliance on AI: An experimental study on the extent and costs of overreliance on AI. *Computers in Human Behavior, 160,* 108352. This experiment shows that simply knowing advice came from an AI can push people to follow it even when it contradicts their own judgment and the evidence in front of them. It sharpens the argument in Section 6.3 that authority is a double-edged tool and that admitting ignorance protects users from harmful overreliance.

Kran, E., Nguyen, H. M., Kundu, A., Jawhar, S., Park, J., & Jurewicz, M. M. (2025). DarkBench: Benchmarking dark patterns in large language models. *Proceedings of the Thirteenth International Conference on Learning Representations (ICLR).* This paper builds a benchmark for detecting manipulative patterns specific to conversational language models, including sycophancy, brand bias, and user-retention pressure. It carries the interface-era taxonomy of Mathur et al. (2019) into the agentic era, giving Section 6.4 a current, AI-native reference point for the dark side of Liking and Authority.

Mathur, A., Acar, G., Friedman, M. J., Lucherini, E., Mayer, J., Chetty, M., & Narayanan, A. (2019). Dark patterns at scale: Findings from a crawl of 11K shopping apps. *Proceedings of the ACM on Human-Computer Interaction, 3*(CSCW), 1–32. This paper provides a rigorous taxonomy of the Dark Patterns mentioned in Section 6.4. It offers modern, digital examples of how principles like Scarcity and Social Proof are weaponized in UI design to trick users, providing a clear contrast to the ethical Guardian Agent approach.

Nass, C., & Moon, Y. (2000). Machines and mindlessness: Social responses to computers. *Journal of Social Issues, 56*(1), 81–103. This paper is crucial for understanding *why* the influence tactics in this chapter work on software. Nass and Moon argue that humans apply social rules (like reciprocity) to computers mindlessly, simply because the computer presents basic social cues. It validates that you don't need to create a complex AI to use these principles; you just need to trigger the right social heuristic.

Rheu, M., Shin, J. Y., Peng, W., & Huh-Yoo, J. (2021). Systematic review: Trust-building factors and implications for conversational agent design. *International Journal of Human–Computer Interaction, 37*(1), 81–96. A systematic review consolidating the factors that drive trust in conversational agents, with personalization and perceived similarity among the strongest. It gives Section 6.2 a research-backed foundation for the claim that mirroring language and referencing shared context build rapport.

Sandoval, E. B., Brandstetter, J., Bartneck, C., & Rötting, M. (2016). Reciprocity in human-robot interaction: A quantitative approach through the prisoner's dilemma and the ultimatum game. *International Journal of Social Robotics, 8*(2), 303–317. This study directly extends the Reciprocity principle into the realm of AI and robotics. It provides empirical evidence that humans act with the same level of reciprocity toward machines as they do toward other humans, confirming that the Gift of Competence strategy is biologically sound.

Strohmetz, D. B., Rind, B., Fisher, R., & Lynn, M. (2002). Sweetening the till: The use of candy to increase restaurant tipping. *Journal of Applied Social Psychology, 32*(2), 300–309. This is the famous candy study detailed in Section 6.1. It isolates the Personalized Gift variable, proving that the highest form of reciprocity is triggered not by the value of the gift, but by the perception that the gift was given specifically and spontaneously to the recipient.

Tufekci, Z. (2015). Algorithmic harms beyond Facebook and Google: Emergent challenges of computational agency. *Colorado Technology Law Journal, 13,* 203. This paper explores the concept of Algorithmic Authority discussed in Section 6.3. Tufekci argues that algorithms have acquired a new form of legitimacy that allows them to structure human attention and decision-making, emphasizing the heavy responsibility designers have when their agents act as gatekeepers of information.

CHAPTER 7

Mechanics of Influence: Social Proof, Scarcity, and Consistency

In the previous chapter, we assembled the first three tools in our psychological toolkit. Reciprocity, Liking, and Authority are the principles of relationship-building. They help our agent show up as a competent, credible, and collaborative partner. They are the tools for making a good first impression and earning the right to be heard. But a good relationship is just the starting point. Our ultimate goal is to build an agent that can effectively help the user accomplish their goals. That process often requires the user to take a specific action or make a commitment.

This is where the final three of Robert Cialdini's principles of influence come into play. Social Proof, Scarcity, and Consistency are the principles of action. They are the psychological levers that help people resolve uncertainty, make decisions, and follow through on their intentions.

If the first three principles are about getting the user to *like* and *trust* the agent, these next three are about getting the user to *agree* with the agent and take the next step. These are arguably the most powerful, and

V. Yocco, *The Persuasive Agent*, Design Thinking,
https://doi.org/10.1007/979-8-8688-3317-5_7

therefore most dangerous, tools in our kit. Wielding them effectively and ethically is the mark of a mature and responsible designer. Let's add them to our workbench.

In this chapter, we will complete the toolkit by examining the Social Proof, Scarcity, and Consistency. We will move beyond the common misconceptions of these terms—where Social Proof is often confused with popularity and Scarcity with pressure—and redefine them as precise engineering patterns. You will learn how to design **Contextual Social Proof** that aids decision-making rather than manufacturing hype. You will learn to frame **Scarcity** as a protective service, turning your agent into a watchdog against loss. Finally, we will use **Consistency** to construct **The Delegation Ladder**, a step-by-step framework for safely increasing an agent's autonomy.

We will conclude with a practical workshop where you will apply these levers to design an ethical, high-conversion interaction flow. Let's get to work.

7.1 Social Proof: Using the Wisdom of the Crowd Responsibly

Imagine you're visiting a new city. You're walking down a street lined with restaurants. One is completely empty. The one next to it has a line of people waiting outside. Which one do you assume is better? Unless you are a very particular kind of contrarian, you will instinctively feel a pull toward the busy restaurant. You don't know anything about the food, the service, or the price. You have a powerful piece of data, however: other people chose this one.

This is the principle of **Social Proof** in action. It states that one of the primary ways we determine what is correct is to find out what other people think is correct. The principle is most powerful in situations of uncertainty. When we are unsure of ourselves, when the situation is ambiguous, we are more likely to look to and accept the actions of others as the right ones. It's a cognitive shortcut. Our System 1 brain uses this heuristic to outsource the hard work of decision-making. If everyone is doing it, it must be the right thing to do.

This heuristic is generally useful. The busy restaurant probably *is* better. But it's also the principle that can lead to bystander apathy, where no one helps a person in distress because they see that no one else is helping. The power of social proof is that it can feel less like influence and more like objective reality.

Social Proof in the Interface Era

The digital world is a giant engine of social proof. It is, without a doubt, the most widely used persuasive principle in modern product design. You see it everywhere:

- **Ratings and Reviews:** Amazon's five-star rating system is perhaps the most famous example. The first thing most of us do on a product page is look for the little yellow stars.
- **Testimonials:** "Customers who bought this also bought..."
- **Subscriber Counts:** "Join 50,000 other developers who get our weekly newsletter."
- **Trending Lists:** Twitter's trending topics or Netflix's Top 10 in Your Country Today row.

These patterns all do one thing: reduce uncertainty and increase confidence. They tell the user, "You are not alone in making this choice. Other people, just like you, have gone before and found value here." It's a powerful way to build Usability Trust and nudge a user toward a specific action.

The Agentic Upgrade: Intelligent, Contextual Social Proof

A simple, static display of a five-star rating is a blunt instrument. An AI agent has the ability to wield social proof with far more nuance, context, and persuasive power. Because an agent understands the user's specific goal, it can provide social proof that is tailored to the user's immediate situation.

Pattern: The People Like You Frame

This is the most common and effective agentic pattern. Instead of showing what everyone thinks, the agent can show what a specific, relevant subgroup thinks. This leverages the principle of Liking (specifically, similarity) to make the social proof exponentially more powerful.

- A **travel agent** could say: "For a family trip to Orlando, most travelers similar to you also booked a rental car with their flight. It saves about 20% on average. Would you like me to look for a package deal?" This is far more persuasive than a generic "People also book rental cars" message. The phrase "travelers similar to you" creates a powerful sense of affinity.
- An **e-commerce agent** helping a user buy a new laptop could say: "Given that you've specified 'video editing' as your primary use case, over 90% of other creative professionals have chosen the model with the upgraded graphics card. They report that it significantly reduces render times." This is a targeted piece of advice from a relevant peer group.

Figure 7-1 is an example of how the people like you frame might be effectively used by a company selling skill building courses to managers who work in tech.

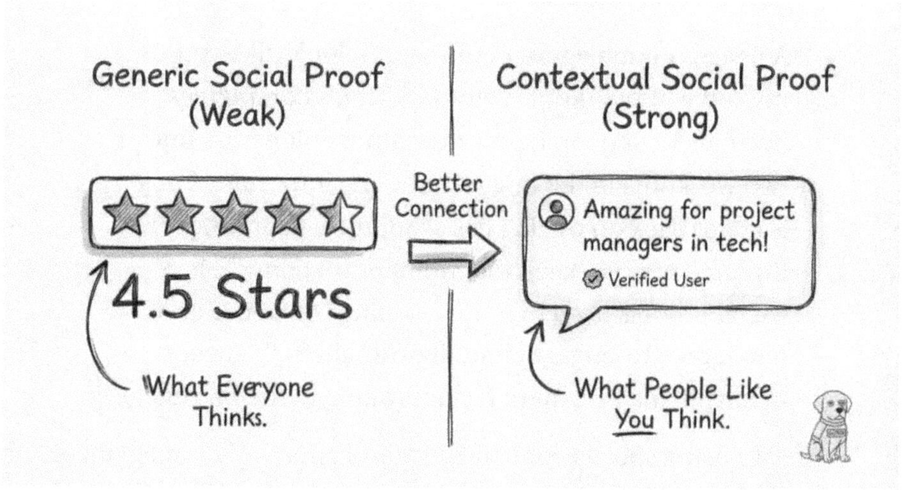

***Figure 7-1.** The Social Proof Spectrum. Generic social proof (popularity) is weak and often ignored. Contextual social proof (relevance) is strong because it helps the user make a specific decision based on peers with similar goals. Image generated by author using Gemini Pro 3.1/Nano Banana 2*

Pattern: The Data-Driven Justification

An agent can use aggregated, anonymous data to justify its suggestions, framing them as the normal or wise choice based on the behavior of others. This is a subtle but powerful way of providing social proof.

- A **project management agent** could suggest: "I see you've set the deadline for this task to tomorrow. Based on similar tasks, teams usually allow three days for a full design review. Would you like to extend the deadline to prevent a potential crunch?" The agent is using social proof (teams usually allow three days) to provide a helpful, data-informed nudge. It feels less like an opinion and more like a piece of expert wisdom derived from the crowd.

- A **fitness coach agent** could say: "It looks like you missed your workout yesterday. That's completely normal! Most people on a new fitness plan miss one or two workouts in the first month. The important thing is to get back on track. Let's schedule a short one for this afternoon to keep the momentum going." Here, social proof is used to reduce feelings of failure and encourage re-engagement. It normalizes the user's behavior, making them feel less alone in their struggle.

The key to using social proof ethically and effectively is relevance. The more targeted, specific, and genuinely helpful the social proof is, the more it will be perceived as a valuable piece of guidance and the less it will feel like a generic marketing ploy.

7.2 Scarcity: The Difference Between Genuine Urgency and a Dark Pattern

The second principle of action is **Scarcity**. The rule is simple: opportunities seem more valuable to us when their availability is limited. This is a powerful motivator because it taps into our deep-seated fear of missing out (FOMO). As something becomes less available, we lose freedom, and we have a visceral, negative reaction to that loss.

Cialdini points to classic examples like the "limited time only" sale or the "only a few left in stock" alert. When the airline booking site tells you, "Only 2 seats left at this price," your System 1 brain feels a jolt of urgency. The rational part of your brain might know it's a sales tactic, but the fear of losing out on the deal often overrides your rational analysis.

Scarcity is incredibly effective at getting people to stop deliberating and start acting. It is also, by far, the most frequently abused and misused principle of influence in the digital world. The landscape is littered with fake countdown timers, false "limited stock" alerts, and high-pressure sales tactics designed to create artificial urgency. This is the fast lane to destroying user trust.

So, how do we use this powerful but dangerous tool ethically? The answer is simple: **only** use scarcity when it is real and when **highlighting it genuinely benefits the user.**

The Ethical Litmus Test for Scarcity

Ask yourself: If I don't show this scarcity information to the user, are they likely to be harmed or disadvantaged later?

- If the answer is yes, then highlighting the scarcity is an act of service.
- If the answer is no, then highlighting the scarcity is an act of manipulation.

Let's apply this. "Only 2 seats left at this price." If the user waits an hour, will that price likely be gone? Yes. Not telling them would be a disservice. This is ethical. A news website with a banner that says, "Special subscription offer ends in 10... 9... 8..." when the same offer will be available tomorrow? That is a manipulative lie.

The Agentic Upgrade: The Proactive Guardian Against Loss

A simple interface can only display scarcity information. An agent, because it understands the user's goals, can take the next step. It can act as a proactive guardian, helping the user *avoid* the negative *consequences* of scarcity. This transforms the agent from a high-pressure salesperson into a helpful lookout who is on the user's side.

In practice, that means tying a piece of real scarcity information directly to the user's stated goal and proposing a contingent action the user can accept or decline. Two short examples show the difference:

- A **travel agent:** "I've found a flight that fits your budget and schedule perfectly. Just so you know, there are only three seats left in that price category. If you'd like, I can put a free 24-hour hold on them for you while you make a final decision. That way, you won't lose the price." The agent has highlighted the scarcity but then immediately offered a helpful, low-commitment solution. It is acting as the user's advocate.
- A **shopping agent:** "I see you're looking at this specific camera model. It's a great choice, but it's currently on backorder at most retailers. I did find one available at Best Buy, but they only have two left in stock. I can't guarantee it will be there for long. Shall I reserve it for in-store pickup for you?" The agent has not only found a scarce item but has also proposed an action to secure it.

Pattern: The Deadline Watchdog

Many real-world opportunities have deadlines. An agent can track these deadlines and proactively alert the user to avoid a loss.

- An agent for a **conference attendee:** "A quick reminder: the early-bird discount for the Design Systems conference ends this Friday. If we register now, we can save you $200. Shall I go ahead and complete the registration with your professional development card?"

- An agent for a **parent:** "Just a heads-up, registration for the summer soccer camp closes tomorrow at 5 PM. I've pre-filled the registration form for Leo with all his information. All you need to do is review it and click 'submit.'"

In all of these cases, the agent is using scarcity ethically. It is identifying a real constraint, connecting it to a user's goal, and proposing a helpful action to avoid a negative outcome. It is not creating false urgency. It is helping the user navigate real-world urgency. This is an act that builds profound trust and demonstrates the unique value of a proactive, goal-aware agent.

Compare that with the same principle turned against the user. Picture a shopping agent that flashes "Only 1 left, and 12 people are viewing this right now" while the warehouse holds 400 units and the viewer count is a random number that changes on every visit. Or a subscription agent whose banner reads "Offer expires in 4:59" and silently resets to five minutes the moment you reload the page. Both invent a constraint that does not exist. The user acts on manufactured panic instead of fact, and the first time they catch the resetting clock or the phantom stock warning, the agent's credibility is gone for good. The tactic in both the ethical and the manipulative version is identical. The difference is whether the scarcity is real, and whether telling the user about it serves the user or the seller.

7.3 Consistency: The Foot-in-the-Door Technique for Building Delegation

The final principle of action is perhaps the most subtle and most important for building long-term delegation trust. The principle of **Consistency** states that once we make a choice or take a stand, we will encounter personal and interpersonal pressure to behave consistently with that commitment.

This desire for consistency is driven by a powerful mental shortcut. Once we have made up our mind about something, we no longer have to expend precious cognitive energy thinking about it. Consistency is a way of automating our behavior. It allows us to be mindless in a way that is usually efficient. People whose beliefs and actions don't align are seen as confused, untrustworthy, or even mentally unstable. Therefore, we have a deep, internal drive to keep our future actions consistent with our past commitments.

Cialdini highlights the classic foot-in-the-door technique. A salesperson doesn't start by asking you to make a huge purchase. They start by asking you to agree to a small, innocuous statement. "You believe it's important to have a safe car for your family, don't you?" Well, yes, of course. That's a small, easy commitment to make. Once you have made that commitment, you are psychologically primed to act consistently with it. You are now more likely to agree to their next, larger request for a set of expensive new tires.

The Agentic Upgrade: The Delegation Ladder

For an AI agent, the principle of consistency is the master key to crossing the Trust Threshold. You do not ask the user for a massive act of delegation on day one. That is the mistake the personal shopping app from Chapter 1 made. Instead, you use the foot-in-the-door technique. You build a **Delegation Ladder**.

The Delegation Ladder is a design strategy where you intentionally scaffold the user's trust over time. You start by asking for a small, low-risk, easily reversible commitment. Once the agent successfully fulfills that small commitment, it has created a positive data point. The user has now made a small, public (even if it's just public to themselves) commitment that this agent is trustworthy. Their desire for consistency will make them slightly more likely to agree to the next, slightly larger request. You guide the user up the ladder of trust, one rung at a time.

Let's design a Delegation Ladder like the one displayed in Figure 7-2 for a new AI-powered email assistant whose ultimate goal is to autonomously manage the user's inbox.

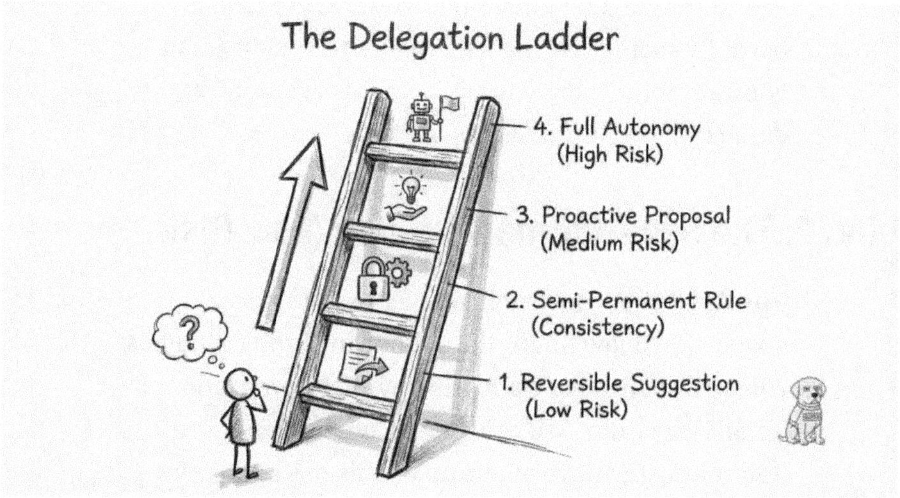

Figure 7-2. *The Delegation Ladder. A framework for safely increasing agent autonomy. Trust is earned by climbing from low-risk, reversible suggestions to high-risk, proactive actions. Image generated by author using Gemini Pro 3.1/Nano Banana 2*

Rung 1: The Small, Reversible Suggestion (Lowest Risk)

- **The Ask:** On day one, the agent doesn't touch anything. It just observes. At the end of the day, it presents a digest: "I noticed you received 15 promotional emails today from retailers. Would you like me to create a 'Promotions' folder and move them there for you?"
- **The Psychology:** This is a perfect foot-in-the-door. The request is small. The action is easily understandable. Most importantly, it is **100% reversible**. If the user doesn't like it, they can just move the emails back. By saying yes, the user makes their first small commitment: "I am willing to let this agent move my emails."

Rung 2: The Semi-Permanent Rule (Low Risk)

- **The Ask:** After the user has accepted the first suggestion a few times, the agent can climb to the next rung. "It seems like we're moving these promotional emails every day. Would you like me to create a rule to automatically move all future emails of this type into your 'Promotions' folder?"
- **The Psychology:** The user has already acted consistently by manually approving the move. This next step is a logical extension of that prior commitment. The risk is still low, as a rule can be easily turned off, but the commitment is slightly larger. The user is now saying, "I trust this agent to handle this specific type of email on its own."

Rung 3: The Proactive, Time-Sensitive Action (Medium Risk)

- **The Ask:** The agent now has a history of competence. It can start to handle more nuanced tasks. It detects an email with a meeting invitation. "I see you've been invited to the 'Project Sync' at 3 PM tomorrow, but your calendar shows you are already booked for a client call. It looks like a conflict. Shall I decline the meeting and suggest an alternative time to the organizer?"
- **The Psychology:** This is a significant step up. The agent is now dealing with the user's schedule and professional reputation. However, it's still an act of co-pilotage. The agent is *proposing* the action, not just taking it. It is showing its work ("It looks like a conflict") and asking for permission. The user's past positive experiences make them more likely to agree. Their commitment is now, "I trust this agent to interact with my colleagues on my behalf."

Rung 4: Full Autonomy (Highest Risk)

- **The Ask:** This is the top of the ladder, a state you might only reach after months of flawless performance. "You've enabled 'Autonomous Scheduling Mode.' An invitation for an internal sync has arrived. I have checked your calendar, confirmed it doesn't conflict with any of your stated priorities for the week, and have accepted it on your behalf. It is now on your calendar."

- **The Psychology:** The agent has now fully crossed the Trust Threshold. The user has delegated a complex task completely. This level of trust was not requested on day one. It was earned, step by step, by a consistent pattern of small, successful commitments.

The Delegation Ladder is one of the most important strategic design patterns for any agentic system. It respects the user's psychology by recognizing that trust is not a switch you flip, but a ladder you climb.

7.4 Ethical Checkpoint: When Social Proof Creates Bias

The principles of action are powerful tools for good, but their potential for misuse is significant. We've already discussed the dark patterns associated with fake scarcity. But the ethical risks of Social Proof and Consistency are more subtle and, in the context of AI, perhaps more dangerous.

An AI agent has the ability to create social proof on a massive, personalized scale. It can surround a user with a seemingly objective reality that is, in fact, a carefully constructed and highly biased point of view.

Imagine a news aggregator agent. If it learns that you tend to click on articles with a certain political slant, it can use that information to create a powerful feedback loop.

1. **Consistency:** It knows you have made past commitments (clicks) to this type of content.
2. **Social Proof:** It then starts to frame its recommendations using social proof. "Here's an article that's trending among people who share your views."

3. **The Result:** The agent has created a filter bubble, an echo chamber where the user is only exposed to information that confirms their existing biases. It feels like the "wisdom of the crowd," but it's actually the wisdom of a very specific, algorithmically selected sub-group. The agent is using these principles not to help the user make a better decision, but to keep them engaged by reinforcing what they already believe.

This is a profound ethical challenge. An AI that has access to vast amounts of data about what "people like you" do has the power to either broaden your horizons or to build the walls of your prison of belief ever higher.

The Ethical Litmus Test for Social Proof

When using social proof, we must ask an additional question:

> *"Is this social proof being used to help the user make a more informed and confident decision among a diverse set of options, or is it being used to narrow the user's perspective and reinforce a pre-existing bias?"*

A responsible agent should use social proof as one data point among many. A travel agent that says, "Most people book a rental car," but also shows you the public transit options is using social proof ethically. An agent that only shows you what "your group" thinks is using it to limit your freedom of choice. This is a subtle but critical distinction.

Table 7-1 shows the contrast between ethical and manipulative patters for the principles of Social Proof, Scarcity, and Consistency.

Table 7-1. *The ethics of action—ethical patterns versus dark patterns for the three principles of action*

Principle	The Guardian Pattern (Ethical)	The Dark Pattern (Manipulative)
Social Proof	**Contextual Wisdom:** Showing data from relevant peers to help the user make an informed choice. ("Professionals like you prefer this model.")	**The False Herd:** promoting a high-margin item by arbitrarily labeling it "Most Popular" or "Trending" to manufacture consensus.
Scarcity	**The Watchdog:** Alerting the user to a real deadline or stock limit to prevent them from losing an opportunity.	**The Fake Countdown:** Using a resetting timer or "High Demand" badge to induce artificial panic and rush a purchase.
Consistency	**The Delegation Ladder:** Asking for small, reversible commitments to slowly build trust and verify alignment.	**The Roach Motel:** Making it easy to sign up (foot-in-the-door) but intentionally difficult to cancel or downgrade.

7.5 Summary

- The final three principles of influence—Social Proof, Scarcity, and Consistency—are the principles of action. They help translate a foundation of trust into user commitment and delegation.
- **Social Proof** is our tendency to rely on the actions of others to resolve our own uncertainty. An agent can use this ethically by providing intelligent, contextual

social proof, such as the **"People Like You" frame** or **Data-Driven Justifications**, to help users make more confident decisions.

- **Scarcity** is our motivation to act when we believe an opportunity is limited. An agent can use this ethically by acting as a **Proactive Guardian Against Loss**, alerting users to *real* scarcity and helping them take action to avoid a negative outcome.
- **Consistency** is our deep-seated desire to have our future actions align with our past commitments. The most important application of this principle is the "foot-in-the-door" technique.
- These principles must be used with extreme care. AI-powered social proof can be used to create biased filter bubbles, and our ethical responsibility is to use it to inform, not to narrow, a user's perspective.

7.6 Chapter 7 Workshop: The Ladder of Trust

We have just added the three most powerful action-oriented principles to our toolkit. These are the levers that move users from passive trust to active commitment. But like any powerful tool, they require skill and care to use correctly. This workshop is your training ground.

You will audit the digital world for these principles, and then you will design a strategic plan to implement them ethically for your own agent. The centerpiece of this activity is designing a Delegation Ladder, which is one of the most critical strategic artifacts you can create for an agentic product.

Part 1: The Scarcity and Social Proof Audit

Let's begin by becoming expert critics. Open a major e-commerce or travel booking website (Amazon, Booking.com, or any airline website are perfect). Your mission is to hunt for examples of Scarcity and Social Proof and apply our ethical litmus tests.

Find one example of each principle on the site and analyze it:

1. **Find an example of Scarcity:** Look for phrases like "Only 2 left in stock," "Sale ends in 2:43:15," or "3 other people are looking at this right now."
   - **The Ethical Litmus Test:** Ask yourself, "If they *didn't* show me this information, could I be disadvantaged later?" Is this real, helpful urgency, or does it feel like a manipulative dark pattern designed to rush your decision? Write down your verdict and a one-sentence justification.

2. **Find an example of Social Proof:** Look for star ratings, customer reviews, "Frequently bought together" modules, or "trending" badges.
   - **The Ethical Litmus Test:** Ask yourself, "Is this social proof helping me make a more informed decision among diverse options, or is it narrowing my perspective?" For example, does the review system allow for easy access to critical (1- and 2-star) reviews, or does it exclusively highlight positive ones? Does it feel like genuine guidance or crowd manipulation?

This exercise will sharpen your senses. You will begin to see the subtle and not-so-subtle ways these principles are used every day, and you will develop a gut feeling for the line between ethical help and manipulative pressure.

Part 2: Designing Your Agent's Delegation Ladder

This is the core of the workshop. Take out the **Agent Persona Canvas** and the **Relationship Blueprint** you have been developing in the previous chapter activities. Think about the ultimate goal of your agent. What is the biggest, most significant task you want users to eventually delegate to it?

Your task is to sketch a four-rung **Delegation Ladder** that will guide the user from a simple first interaction to that ultimate goal. Be specific. For each rung, define the agent's ask and explain the psychology behind it.

Example Project: An AI agent for a grocery shopping app.

- **Ultimate Goal:** For the agent to autonomously generate the user's weekly shopping list based on their past purchases, meal plans, and what's on sale, and then place the order for delivery.

Your Delegation Ladder:

- **Rung 1: The Small, Reversible Suggestion (The Foot-in-the-Door)**
  - **The Ask:** *What is the first, tiny, zero-risk request your agent makes?* Example: As you add milk to your cart, the agent says: "I notice you buy this brand of milk every week. Would you like me to add it to a 'Weekly Staples' list for you?"
  - **The Psychology:** *[Explain why it's low-risk.]* Example: This is a simple categorization task. It is 100% reversible (the user can remove it). Saying yes is a small commitment to the idea that the agent can help organize things.

- **Rung 2: The Semi-Permanent Rule (Building Consistency)**

- **The Ask:** *[What is the next logical step that builds on the first commitment?]* Example: After a few weeks, the agent says: "We've been adding the same 5 items to your 'Weekly Staples' list. Would you like me to automatically add these 5 items to your cart for you at the start of every shopping session?"
- **The Psychology:** *[Explain how it leverages the user's past actions.]* Example: The user has already consistently identified these items as staples. This automates that commitment. The risk is still low, as items can be easily removed from the cart.

- **Rung 3: The Proactive, Contextual Action (Medium Risk)**
  - **The Ask:** *[What is a more complex, proactive suggestion that shows deeper understanding?]* Example: The agent sees you've added pasta and ground beef to your cart. It says: "It looks like you might be making bolognese. The recipe you saved last month also called for crushed tomatoes and onions. You don't have those in your cart. Shall I add them?"
  - **The Psychology:** *[Explain why this requires more trust.]* Example: The agent is now making inferences about the user's intent. It's a significant step up in competence. A successful suggestion here builds a huge amount of trust in the agent's "judgment."

- **Rung 4: Full Autonomy (The Ultimate Delegation)**
  - **The Ask:** *[What is the final, top-of-the-ladder autonomous action?]* Example: The agent sends a push notification on Sunday morning: "Based on your staples and meal plan, I've prepared your

shopping list for the week. The total is $87.50, which is about average for you. Would you like me to place the order with your saved payment method for delivery this afternoon?"

- **The Psychology:** *[Explain why this is the culmination of the process.]* Example: This is full delegation. The user only has to provide the final "yes." This level of trust was earned rung by rung, through a series of successful, consistent, and low-risk interactions.

Part 3: The Socially Aware Action Enabler

Now, let's layer in the other two principles to support the ladder you just built. For your agent, brainstorm one specific pattern for each.

1. **Your Social Proof Strategy:** How can you use the "wisdom of the crowd" to help a user feel more confident in one of the rungs on your ladder?
    - **Example (for Rung 3 above):** When suggesting the crushed tomatoes, the agent could add: "Over 90% of people who make this recipe also add a bit of parmesan cheese at the end. It's not on your list, but I thought you might like to know!"
2. **Your Scarcity Strategy:** How can you use real, ethical scarcity to act as a Guardian Against Loss for your user?
    - **Example (for Rung 4 above):** When presenting the final cart, the agent could say: "Just a heads-up, the organic strawberries you love are on a special 2-for-1 sale, but the offer ends today. The list I've prepared includes two cartons to take advantage of the deal."

By completing this workshop, you have moved beyond designing isolated features and have started to think like a true agent designer. You have created a long-term strategic plan for earning user trust and a set of tactical interventions for guiding user action. This combination of a long-term strategy (the Delegation Ladder) and tactical interventions (Social Proof, Scarcity) is the core of designing effective, ethical, and ultimately successful agentic experiences

7.7 References and Further Reading

Asch, S. E. (1951). Effects of group pressure upon the modification and distortion of judgments. In H. Guetzkow (Ed.), *Groups, leadership and men* (pp. 177–190). Carnegie Press. This is the classic conformity study mentioned in Section 7.1. Asch demonstrated that individuals will often ignore the evidence of their own senses (the length of a line) to align with a group consensus. It provides the foundational evidence for Social Proof: when in doubt, humans look to others to determine what is true.

Chu, H., & Liu, S. (2024). Can AI tell good stories? Narrative transportation and persuasion with ChatGPT. *Journal of Communication, 74*, 347–358. This study extends the Social Proof discussion in Section 7.1 into the domain of narrative persuasion. Chu and Liu find that ChatGPT-generated stories transport readers into a narrative world with an effect comparable to human-written stories, and that this transportation predicts persuasion. It is a reminder that an agent's storytelling, not just its stated facts, can move a user toward a decision.

Cialdini, R. B. (1984). *Influence: The psychology of persuasion.* Harper Business. Continuing from Chapter 6, this text remains the primary framework for the principles of Social Proof, Scarcity, and Consistency. It details the mechanisms of "Click, Whirr" behavior that allow these principles to act as shortcuts for decision-making.

Freedman, J. L., & Fraser, S. C. (1966). Compliance without pressure: The foot-in-the-door technique. *Journal of Personality and Social Psychology, 4*(2), 195–202. This is the definitive study for the "Delegation Ladder" concept in Section 7.3. The authors proved that getting a person to agree to a small, trivial request (putting a small sign in their window) drastically increased the likelihood they would agree to a massive request later (putting a giant billboard on their lawn), validating the principle of Consistency.

Kahneman, D., & Tversky, A. (1979). Prospect theory: An analysis of decision under risk. *Econometrica, 47*(2), 263–291. This Nobel Prize-winning paper introduces Prospect Theory and the concept of "Loss Aversion." It explains the biological mechanism behind Scarcity (Section 7.2): humans are psychologically wired to fear loss twice as much as they value an equivalent gain, which is why "avoiding a missed opportunity" is such a powerful motivator.

Lim, S., & Schmälzle, R. (2024). The effect of source disclosure on evaluation of AI-generated messages. *Computers in Human Behavior: Artificial Humans, 2,* 100058. This paper supports the ethical litmus test proposed in Section 7.4. Lim and Schmälzle show that disclosing an AI source changes how a message is evaluated and how persuasive it is judged to be, providing direct evidence for the argument that transparency about who or what is speaking is part of ethical persuasion design.

Ma, S., Wang, X., Lei, Y., Shi, C., Yin, M., & Ma, X. (2024). "Are you really sure?" Understanding the effects of human self-confidence calibration in AI-assisted decision making. *Proceedings of the 2024 CHI Conference on Human Factors in Computing Systems (CHI '24),* Article 840. This CHI paper grounds the Delegation Ladder introduced in Section 7.3. Ma and colleagues find that calibrating a user's confidence against a system's actual accuracy improves reliance decisions, supporting the case for building trust step by step rather than asking for full delegation on day one.

Mathur, A., Acar, G., Friedman, M. J., Lucherini, E., Mayer, J., Chetty, M., & Narayanan, A. (2019). Dark patterns at scale: Findings from a crawl of 11K shopping apps. *Proceedings of the ACM on Human-Computer Interaction, 3*(CSCW), 1–32. Essential for Section 7.4, this paper analyzes how Scarcity and Social Proof are weaponized in modern apps. It documents the prevalence of "activity messages" (fake social proof) and "countdown timers" (fake scarcity), providing the counter-examples to your ethical "Guardian Agent" approach.

Okoso, A., Yang, M., & Baba, Y. (2025). Do expressions change decisions? Exploring the impact of AI's explanation tone on decision-making. *Proceedings of the 2025 CHI Conference on Human Factors in Computing Systems (CHI '25)*, Article 824. Relevant to the Data-Driven Justification pattern in Section 7.1. Okoso, Yang, and Baba demonstrate that the tone of an AI's explanation, independent of its content, measurably changes user decisions. It reinforces the chapter's claim that how an agent frames a nudge carries as much weight as the nudge itself.

Panda, S. K., & Schecter, A. (2024). Advice utilization in combined human-algorithm decision-making: An analysis of preferences and behaviors. *Journal of the Association for Information Systems, 25*(6), 1439–1465. This paper supports the Data-Driven Justification pattern discussed in Section 7.1. Panda and Schecter examine when people accept versus discount algorithmic advice, offering evidence for designing agent suggestions that are more likely to be adopted rather than dismissed.

Pariser, E. (2011). *The filter bubble: What the Internet is hiding from you.* Penguin Press. This book is the key reference for the ethical risks discussed in Section 7.4 regarding Social Proof and bias. Pariser defines the "Filter Bubble"—the intellectual isolation that occurs when algorithms selectively guess what information a user would like to see based on past behavior—reinforcing the danger of using Consistency and Social Proof to narrow a user's worldview.

Rogiers, A., Noels, S., Buyl, M., & De Bie, T. (2024). Persuasion with large language models: A survey. *arXiv preprint* *arXiv:2411.06837*. This survey is the broadest reference point for the chapter's central concern. Rogiers and colleagues review the growing body of work showing that LLM systems have already achieved human-level or greater persuasiveness across politics, marketing, and public health, and catalog the ethical risks, including manipulation and the erosion of user autonomy, that motivate the Guardian Agent framing used throughout this chapter.

Schoenegger, P., Costello, T. H., Pennycook, G., & Rand, D. G. (2025). Large language models are more persuasive than incentivized human persuaders. *arXiv preprint* *arXiv:2505.09662*. This study raises the stakes for everything in this chapter. Schoenegger and colleagues find that LLMs, even without any special optimization for persuasion, outperform paid human persuaders in changing people's minds. It is the clearest evidence that the ethical guardrails proposed here are not precautionary but urgent.

Teeny, J. D., & Matz, S. C. (2024). We need to understand "when," not "if," generative AI can enhance personalized persuasion. *Proceedings of the National Academy of Sciences, 121*, e2418005121. This PNAS paper reframes the question at the heart of Section 7.1's Agentic Upgrade. Teeny and Matz argue that generative AI's persuasive power is not a fixed property but depends heavily on context, audience, and design choices, echoing this chapter's distinction between contextual, ethical social proof and generic, manipulative social proof.

Worchel, S., Lee, J., & Adewole, A. (1975). Effects of supply and demand on ratings of object value. *Journal of Personality and Social Psychology, 32*(5), 906–914. This is the "Cookie Jar" study that validates the Scarcity principle discussed in Section 7.2. The researchers found that the exact same cookie was rated as more desirable and valuable when it was in a jar of two cookies (scarce) than when it was in a jar of ten (abundant), proving that availability directly alters our perception of quality.

CHAPTER 8

The Transparency Paradox: Explainability and Calibrated Trust

In Chapter 7, we built a Delegation Ladder using Social Proof, Scarcity, and Consistency to guide users toward action. We gave your agent the psychological tools to secure agreement. Now, we face a new problem: when the agent takes action, how much should it explain?

Product teams debate artificial intelligence transparency. The standard argument assumes these systems operate inside hidden boxes. To build trust, we must open the lid for our users to see what's inside. We assume explaining the math behind a recommendation leads directly to user trust. Many teams treat a complicated explanation as a universal fix.

This assumption is often wrong. Supplying excess data that overwhelmed users can decrease trust, reduce adoption, and break the user experience. This is the Transparency Paradox. In this chapter, we map exactly when to explain an agent's reasoning and when to stay silent. We examine the psychological phenomenon of algorithm aversion. You will learn to use the Transparency Triage Framework to evaluate risk

V. Yocco, *The Persuasive Agent*, Design Thinking,
https://doi.org/10.1007/979-8-8688-3317-5_8

and unexpectedness. We then translate these decisions into concrete UI patterns, like an expandable link on a travel booking screen, to achieve calibrated trust.

8.1 Surprising Research on AI Disclosure

For the past several years, a single word has dominated nearly every conversation about the ethics and user experience of artificial intelligence: "transparency." The argument, on its face, is simple and overwhelmingly compelling. AI systems are complex, opaque "black boxes." To make them safer, fairer, and more trustworthy, we must pry open those boxes and show users how they work. The more we explain, the more users will understand. The more they understand, the more they will trust. Transparency, we have been told, is the silver bullet for building user trust.

It's a beautiful idea. It appeals to our sense of intellectual honesty and user empowerment. And in many cases, it is absolutely correct. The problem is that in a surprising number of other cases, it is completely wrong. Sometimes, more transparency can actually decrease user trust, reduce adoption, and make an AI system less effective. This is the Transparency Paradox, and understanding its nuances is one of the most critical skills for any designer of agentic systems.

I first encountered this contradiction while reviewing a fascinating piece of research from a few years ago (Reis, Reis, & Kunde, 2024). A team of researchers wanted to see how disclosing the use of an AI would affect people's perception of medical advice. They presented participants with a doctor's diagnosis of a skin lesion.

One group was simply told, "Here is the diagnosis from the doctor."

Another group was told, "Here is the diagnosis from the doctor, who was assisted by a highly accurate AI system."

Logically, the second group should have more confidence. The diagnosis was backed by a human expert and a powerful machine. But that's not what happened. The participants who were told an AI was involved rated the diagnosis as less trustworthy. Less expectedly, they rated the human doctor who used the AI as being less competent than the doctor who worked alone. The very act of revealing the AI's involvement poisoned the well. The researchers called this "algorithm aversion." A large 2025 replication found the same penalty lands on the physician's own reputation as much as on the AI: simply mentioning AI use, even for scheduling, made patients rate the doctor as less competent and less trustworthy (Reis, Reis, & Kunde, 2025).

This isn't an isolated finding. Study after study has shown that our relationship with transparency is far more complex than we assume. Another study found that when a task is perceived as highly subjective or intuitive (like recommending a movie or hiring a job candidate), people tend to trust a human expert's "gut feeling" more than an algorithm's data-driven conclusion, even when the algorithm is demonstrably more accurate. We seem to have an innate bias that for tasks involving human nuance, algorithms are somehow "cheating" or missing the point.

What's going on here? The simple "more transparency equals more trust" model fails to account for the powerful psychological noise we discussed in Chapter 3. As a reminder, psychological noise is the set of internal states a person brings to an interaction, their biases, fears, and mental models, that color how they read a message before they ever weigh its content. When we introduce an AI into the equation, we trigger a whole host of pre-existing biases, fears, and mental models in the user's mind.

- **Fear of the Unknown:** For many, AI is still a scary, sci-fi concept associated with job losses and robot uprisings. Revealing its presence can trigger anxiety rather than confidence.

- **Aversion to Imperfection:** We know humans make mistakes. We are forgiving of human error. But we often hold machines to a standard of perfection. When an AI makes a mistake, it can feel more jarring and trust-destroying than when a human makes the exact same error.

- **The Weirdness Factor:** Sometimes, the AI's reasoning, even if accurate, is just plain weird. A recommendation engine might correctly identify that people who buy hot sauce and classical music also tend to buy a specific brand of cat litter. This correlation may be statistically sound, but it's inscrutable to a human. Exposing this reasoning doesn't create clarity; it creates a feeling that the machine is a strange, alien intelligence.

This doesn't mean transparency is bad. It means that transparency is not a simple on/off switch. It is a highly sensitive design element that must be deployed with skill, precision, and a deep understanding of the user's psychological context. Pouring a firehose of raw technical data on a user is not transparency; it's a form of cognitive waterboarding. The goal is not perfect, absolute transparency. The goal is something much more useful: calibrated trust.

8.2 The Goal Is Calibration, Not Absolute Trust

One of the most dangerous myths in our field is the idea that the goal of a good user experience is to maximize user trust. We should not be trying to get users to trust our agents absolutely and uncritically. That is not a design goal; it is a recipe for disaster. Afroogh, Akbari, Malone, Kargar, and Alambeigi (2024) catalog the specific trust-breakers and trust-makers behind exactly this failure mode.

Blind trust in an AI is just as dangerous as blind distrust. A user who unquestioningly accepts every suggestion from a coding assistant without reviewing the code is going to ship bugs. A user who follows a GPS navigation agent's directions off a collapsed bridge is a victim of over-trust. A user who delegates their entire financial portfolio to an untested robo-advisor without understanding the risks is setting themselves up for ruin.

The goal is not to maximize trust. The goal is to calibrate it.

Calibrated Trust means the user has a realistic and accurate mental model of the agent's capabilities: they understand what it is good at, what it is bad at, and where its knowledge ends. A well-calibrated user knows when to rely on the agent and when to be skeptical. They know when to delegate and when to double-check. They treat the agent not as an infallible oracle, but as a competent but imperfect partner.

I have a personally calibrated level of trust with my car's navigation system. I trust it implicitly to get me 95% of the way to a new destination. I do *not* trust its final "turn left into the parking lot" instruction, because it has been hilariously wrong about the specific entrances to buildings on multiple occasions. My trust is calibrated. I know its strengths (long-distance routing) and its weaknesses (fine-grained local knowledge). This calibration makes the tool incredibly useful to me. If I trusted it blindly, I'd be frustrated every time I ended up in a loading dock or at the bottom of a river. If I distrusted it completely, I'd lose the benefit of its core function.

Everything we do in designing for transparency should be in service of this one goal: helping the user build an accurate mental model of the agent. This means we must be as honest about our agent's weaknesses as we are about its strengths. This is a direct application of the principle of Authority we discussed in Chapter 6. An agent that confidently admits its limitations is building calibrated trust.

This shifts our entire perspective. The question is no longer, "How much can we explain?" The question becomes, "What information does the user need right now to make an informed decision about whether or not to trust this specific suggestion?" The answer is rarely a full technical schematic. The answer is usually a small, targeted, and easily digestible piece of information that addresses the user's most likely uncertainty at that precise moment.

8.3 Transparency Triage Framework: What, When, and How to Explain

If we can't (and shouldn't) explain everything all the time, how do we decide what to explain and when? We need a mental model, a decision-making tool, the **Transparency Triage Framework**.

In a hospital emergency room, a triage nurse doesn't treat every patient the same. They rapidly assess the situation and prioritize based on severity. We need to do the same for explainability. For any given action or suggestion our agent makes, we need to quickly assess the situation and provide the appropriate level of explanation.

The framework is based on two simple axes: **Risk** and **Unexpectedness**.

- **Risk:** What are the consequences to the user if the agent gets this wrong? A low-risk action is easily reversible and has minimal cost (e.g., suggesting a song). A high-risk action is costly, hard to reverse, and could have serious negative consequences (e.g., executing a stock trade).
- **Unexpectedness:** How surprising or counter-intuitive is the agent's suggestion? A highly expected suggestion aligns perfectly with the user's mental model (e.g., suggesting the user's favorite pizza place). A highly unexpected suggestion violates their mental model (e.g., suggesting a sushi place when the user has only ever ordered pizza). Figure 8-1 plots these two axes against each other, giving us a four-zone matrix. Each zone calls for a different level of explanation, which we walk through next.

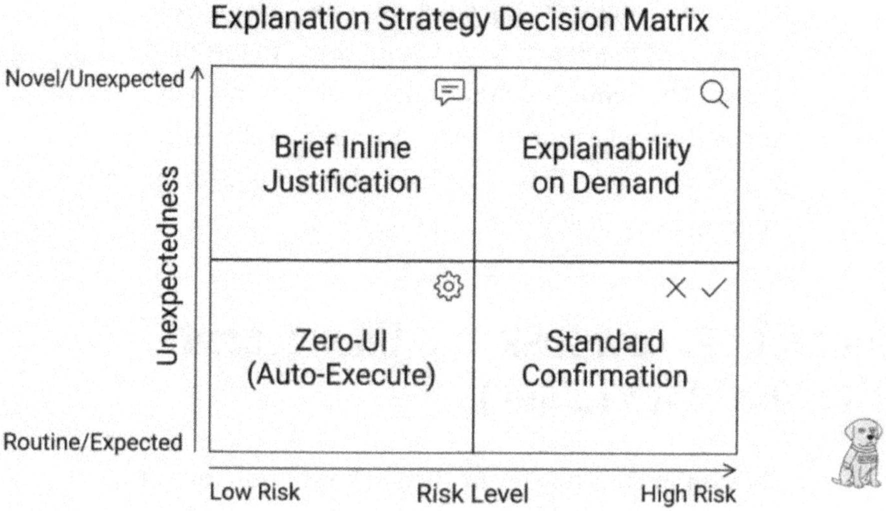

Figure 8-1. *The Transparency Triage Matrix. A framework for determining the exact level of explainability an agent should provide by plotting the risk of an action against how unexpected that action is to the user. Image generated by author using Gemini Pro 3.1/Nano Banana 2*

By plotting our agent's action on this 2x2 grid, we can determine the right level of transparency.

Quadrant 1: Low Risk and Expected (The "Just Do It" Zone)

- **Example:** A music app you've used for years (Spotify) suggests a playlist of your favorite artists.
- **User's Mental State:** Yep, that makes sense.
- **Required Explanation: None.** This is the "don't make me think" zone. Providing an explanation here would be actively annoying. It would be like a barista handing

you your usual morning coffee and then giving you a five-minute lecture on why they chose that specific blend. The agent should just perform the action or make the suggestion cleanly and confidently. Over-explaining the obvious is a form of psychological noise that signals a lack of social awareness.

Quadrant 2: Low Risk and Unexpected (The Huh? Why? Zone)

- **Example:** Your music app, which knows you love indie rock, suddenly suggests a classical cello concerto.
- **User's Mental State:** That's weird. Why would it suggest that?
- **Required Explanation: A brief, simple justification.** The user's mental model has been violated, and you need to provide a short explanation to repair it. The goal is to make the unexpected feel reasonable.
  - *Good Explanation:* "You might like this because the chord progressions are similar to your favorite band, The National." This provides a single, easy-to-understand data point that connects the unknown to the known. It gives the user a reason to trust the suggestion and potentially expand their horizons.
  - *Bad Explanation:* "Our collaborative filtering model indicates a 78% probability of preference based on cluster analysis of 10,000 similar users." This is technically accurate but psychologically useless. It doesn't help the user make a decision; it just confuses them.

Quadrant 3: High Risk and Expected (The Just Confirm Zone)

- **Example:** You tell your financial agent to buy 10 shares of Apple stock, and it prepares the order.
- **User's Mental State:** "This is important. I need to make sure it's exactly right."
- **Required Explanation: A clear, concise confirmation of the critical details.** This is not the time for storytelling. This is the time for precision. The user doesn't need to know *why* you are buying the stock; they just told you to. They need absolute certainty about the *what.*
    - *Good Explanation:* A standard confirmation dialog: "You are about to place a market order for 10 shares of AAPL at about $175.20 per share, roughly $1,752.00 in total. The order will execute right away at the best available price, which can move a little before it fills. Please confirm."
    - *Bad Explanation:* "Based on your request, we're executing a trade for a well-known tech company." This is dangerously vague and would destroy any trust the user has.

Quadrant 4: High Risk and Unexpected (The Whoa, Hold On! Zone)

- **Example:** You ask your financial agent for a safe investment, and it suggests putting 20% of your portfolio into a highly volatile cryptocurrency.
- **User's Mental State:** "That seems like a terrible idea. This is important, and I completely disagree. I need to understand everything."
- **Required Explanation: Full, layered, and on-demand explainability.** This is the highest level of triage. The agent has made a high-stakes suggestion that directly contradicts the user's expectations. To have any hope of gaining the user's trust, it must be prepared to show all of its work. The initial explanation should be a clear, concise summary, but it must be accompanied by an easy way to get more detail.
    - *Good Explanation:* "I know this seems unusual, but I'm suggesting a small allocation to CryptoFund X for a few reasons. While volatile, its performance has been historically uncorrelated with the stock market, which could add diversification to your portfolio and potentially hedge against market downturns. For more details, you can view the [Backtesting Report], the [Risk Analysis], and the [Fund Prospectus]."

    This provides a clear top-level reason and then offers multiple layers of deeper information for the user to explore based on their own level of interest. This is explainability on demand.

The Transparency Triage framework is a dynamic tool. It forces you to stop thinking about explainability as a feature and start thinking of it as a service you provide to the user, tailored to their specific needs in a given moment. Next, we apply these triage decisions to actual user interfaces using a concept called explainability on demand.

8.4 Designing for Explainability on Demand

The most sophisticated and user-respecting way to handle transparency, especially in that critical fourth quadrant, is to design for **explainability on demand**. This is an interaction design philosophy that balances the need for a clean, simple interface with the user's need for deeper understanding. This mirrors a broader move the field has already documented, from static explainability toward interactive, user-directed engagement with an AI's reasoning (Raees, Meijerink, Lykourentzou, Khan, & Papangelis, 2024).

The core principle is this: Present the simplest possible explanation by default, but make it effortless for the user to progressively disclose more detail if they want it.

This approach respects the user's cognitive load. It doesn't overwhelm the majority of users who are satisfied with the top-level explanation, but it empowers the curious or skeptical user to go as deep as they want.

Let's look at some common design patterns for this.

Pattern: Progressive Disclosure: The Expandable Why? Link

This is the most common and effective pattern. The agent makes its suggestion, provides a one-sentence summary of its reasoning, and includes a small link or button that says "Why this suggestion?" or "How was this calculated?"

- **Example:** A product recommendation agent shows you a new camera Sony Alpha a7 IV.

  I'm recommending this because it's highly rated for the type of low-light photography you do.

  [Why this recommendation?]

Clicking the link doesn't take the user to a new page. Instead, it gracefully expands a small section in-line, revealing the next layer of detail.

- Expert Reviews: Rated 4.8/5 stars for low-light performance by DPReview.
- Peer Activity: Three other professional photographers you follow on this platform have recently purchased this model.
- Feature Match: Its sensor size is ideal for the nighttime cityscapes you often shoot.

This is beautiful because it's optional. The user pulls the information they need, rather than having it pushed on them. Figure 8-2 shows how this might look in the context of an Agent charged with booking business travel for a team member.

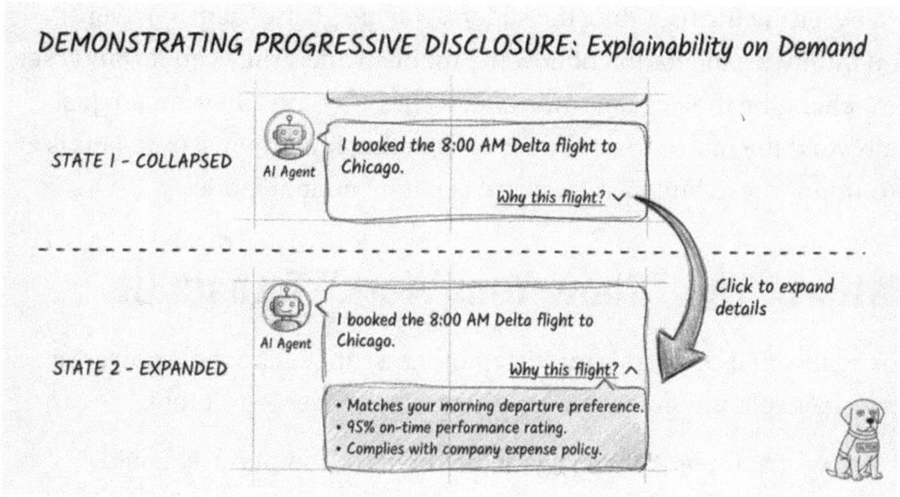

***Figure 8-2.** Explainability on Demand. This wireframe demonstrates progressive disclosure. A user clicks a link to view the agent's reasoning without cluttering the primary interface. Image generated by author using Gemini Pro 3.1/Nano Banana 2*

Pattern: The Interactive Slider or Control

Sometimes, the best way to explain a system is to let the user play with the variables. This is common in financial or scientific modeling tools.

- **Example:** An agent that helps you plan for retirement might present a graph showing your projected savings. Instead of a static report, key variables are presented as sliders. "Based on your current savings rate of 10% and an assumed market return of 7%, you are on track to retire at age 65 with $1.2 million."

  Your Savings Rate: [Slider control, currently at 10%]

  Assumed Market Return: [Slider control, currently at 7%]

By letting the user drag the sliders, you are giving them a powerful and intuitive explanation of how the model works. They can instantly see how changing their savings rate affects the outcome. They are not just being told the reasoning; they are discovering it for themselves. This is a profound way to build a robust and accurate mental model.

Pattern: The "Show Your Work" Transcript

For agents that perform a multistep process, one of the most powerful forms of explanation is to simply show a log of the steps it took.

- Example: A travel agent that books a complex trip could provide a simple summary: "Okay, your trip to London is booked!" followed by an expandable link: [View booking process log].
- Clicking it would reveal a step-by-step transcript:
    1. *Searching for flights on your preferred airline...*
    2. *Found a direct flight matching your schedule.*
    3. *Cross-referencing with your company's travel policy... Approved.*
    4. *Applying your frequent flyer number...*
    5. *Booking seat 14C as requested...*
    6. *Flight confirmed. Sending confirmation to your email and calendar.*

This detailed log is immensely reassuring. It proves the agent didn't just magically produce a result; it followed a logical, competent, and correct process. It is the ultimate expression of "showing your work."

Designing for explainability on demand is the key to resolving the Transparency Paradox. It allows you to build a system that *is* transparent, without forcing that transparency on the user in a way that creates cognitive overload and psychological noise. It respects the user's intelligence and autonomy, giving them the tools they need to build a properly calibrated and resilient sense of trust in your agent. Table 8-1 gathers the patterns from this section into one reference, matching each pattern to its triage quadrant, its risk and unexpectedness profile, the UI strategy behind it, and a worked example you can adapt.

A sophisticated approach to transparency is a core component of designing a trustworthy agent that respects the user's intelligence and empowers them to make informed decisions about delegation. In Chapter 9, we turn our attention to the cognitive biases users bring to the table, and how to address these in our agents' design.

Table 8-1. *The explainability design pattern library*

UI Pattern Name	Triage Quadrant	Risk Level	Unexpectedness	UI Execution Strategy	Practical Example
Zero-UI (The Just Do It Zone)	Quadrant 1	Low	Expected	No explanation provided; execute the action or suggestion cleanly.	Suggesting a playlist of favorite artists.
The Brief Inline Justification	Quadrant 2	Low	Unexpected	A short, single-sentence justification connecting the unknown to the known.	Explaining a classical music suggestion by noting chord similarities to a favorite rock band.

(*continued*)

Table 8-1. *(continued)*

UI Pattern Name	Triage Quadrant	Risk Level	Unexpectedness	UI Execution Strategy	Practical Example
The Standard Confirmation Dialog	Quadrant 3	High	Expected	A clear, concise confirmation of critical details requiring explicit user approval.	A prompt to confirm the purchase of 10 shares of a requested stock.
The Expandable Why? Link	Quadrant 4	High	Unexpected	A top-level summary with a link that expands in-line to reveal deeper data layers.	Recommending a camera with an expandable section detailing expert reviews and peer activity.
The Interactive Slider	Quadrant 4	High	Unexpected	Presenting key variables as interactive controls so users can discover the reasoning by manipulating the model.	Sliders for savings rate and market return in a retirement planning agent.

(continued)

Table 8-1. (*continued*)

UI Pattern Name	Triage Quadrant	Risk Level	Unexpectedness	UI Execution Strategy	Practical Example
The "Show Your Work" Transcript	Quadrant 4	High	Unexpected	Providing an expandable link to a step-by-step log of the agent's automated process.	A chronological log showing how a travel agent cross-referenced policies to book a flight.

8.5 Summary

- The common wisdom that "more transparency equals more trust" is often wrong. The **Transparency Paradox** describes how disclosing the use of AI can sometimes decrease user trust due to psychological noise like algorithm aversion and fear of the unknown.
- The goal of explainability is not to maximize trust; it is to **calibrate** it. We want users to have an accurate and realistic mental model of our agent's strengths and weaknesses so they know when to rely on it and when to be skeptical.
- The **Transparency Triage Framework** is a practical tool for providing the right level of explanation at the right time. By assessing the **Risk** and **Unexpectedness** of an agent's action, we can decide whether to provide no explanation, a brief justification, a detailed confirmation, or a full, layered explanation.

- **Explainability on Demand** is the most effective design philosophy for handling transparency. It involves presenting the simplest explanation by default but making it effortless for users to progressively disclose more detail through patterns like expandable "Why?" links, interactive controls, and "Show Your Work" transcripts.
- A sophisticated approach to transparency is not a "nice-to-have" feature. It is a core component of designing a trustworthy agent that respects the user's intelligence and empowers them to make informed decisions about delegation

8.6 Chapter 8 Workshop: Triage and Transparency Blueprint

In this chapter, we dismantled the simplistic idea that "more transparency is always better" and replaced it with a more sophisticated goal: **calibrated trust**. We introduced the Transparency Triage Framework as our tool for achieving that goal. Now, it's time to put that tool to work.

This workshop will train you to think like a triage nurse for information. You will practice rapidly assessing an agent's actions and then designing the appropriate level of explanation. This is a crucial skill for preventing cognitive overload wand building a resilient, trusting relationship with your user.

Part 1: Triage Drill

Let's build some muscle memory. Below are four scenarios involving different AI agents. Your task is to place each one into the correct quadrant of the **Transparency Triage Framework (Risk vs. Unexpectedness)**. For each, identify the quadrant and write a one-sentence justification for your choice.

- **Scenario A:** You've asked your AI coding assistant to generate a simple "for loop" to iterate through an array. It instantly provides the standard, correct boilerplate code.
  - **Quadrant:**
  - **Justification:**
- **Scenario B:** Your Netflix-style streaming agent, which knows you watch a lot of dark science fiction, recommends a lighthearted British baking show from the 1990s.
  - **Quadrant:**
  - **Justification:**
- **Scenario C:** Your smart home agent, which you've configured to always lock the doors at 11 PM, sends you a notification at 11:01 PM that says, "All doors have been locked for the night."
  - **Quadrant:**
  - **Justification:**
- **Scenario D:** Your personal finance agent sends you an alert: "Based on a new fraud detection algorithm, I have flagged your recent $1,200 payment to 'Vacation Rentals LLC' as potentially suspicious and have temporarily frozen your credit card."
  - **Quadrant:**
  - **Justification:**

This drill trains you to instinctively analyze an agent's action through the two critical lenses of risk and surprise. It's the first step in moving from a one-size-fits-all approach to a nuanced, context-aware strategy for explainability.

(Answers for your reference: A=Q1, B=Q2, C=Q3, D=Q4)

Part 2: Designing for the "Whoa, Hold On!" Moment

Scenario D is a classic Quadrant 4 problem: it is both extremely high-risk (my card is frozen!) and highly unexpected (I thought that payment was legitimate!). This is the moment where an agent's ability to explain itself is most critical. A bad explanation will cause panic and destroy trust. A good one can turn a moment of fear into a moment of profound gratitude and trust-building.

Your task is to design the **Explainability on Demand** for this scenario.

1. **The Top-Level Explanation:** Write the exact, initial alert message the user would see. It needs to be calm, clear, and provide the most critical information up front without causing undue alarm. It should also clearly signal that more information is available. (Hint: The example in the scenario is a good start, but can you improve it?)

2. **The First Layer of Detail:** The user has clicked the "[Why was this flagged?]" link in your alert. Now, design the next layer of information. What are the 2–3 key, easy-to-understand data points that justify the agent's suspicion? You don't need to be a fraud detection expert; imagine a plausible, user-friendly explanation.

- *Example bullet points:*
  - "This merchant has no transaction history with you."
  - "The merchant's website was created only 48 hours ago."
  - "Several other users have flagged transactions from this merchant as fraudulent in the last 24 hours."

3. **The "Show Your Work" Pattern:** What would be the final, deepest layer of explanation? What could the user access if they are still skeptical? Brainstorm one or two "Show Your Work" options the agent could provide.
   - *Example ideas:* A link to a log of the agent's automated security checks, a button to immediately initiate a chat with a human fraud specialist, a map showing the unusual geographic location of the transaction.

This exercise shows how to layer information effectively. You provide a calm summary, an accessible justification, and an escape hatch to deep evidence. This respects the user's intelligence and gives them control over how much detail they engage with. Below is a worked example for this part of the activity.

A Sample Answer

Here is one workable version. Yours can differ in voice, and the point is to see how the three layers stack.

1. **The top-level explanation (the first thing the user sees):**

I paused your card for a moment. A $1,200 payment to "Vacation Rentals LLC" looks different from your usual activity, so I placed a temporary hold while we confirm it was you. Nothing has been lost, and you can lift the hold in two taps. [Why was this flagged?] [This was me, unfreeze now]

It stays calm, leads with the reassurance that nothing is lost, gives the user an immediate way out, and signals that a reason is one tap away.

2. **The first layer of detail (after the user taps "Why was this flagged?"):**

   - *This payee has no prior history on your account.*
   - *The charge posted at 3:14 a.m. local time, outside your normal spending hours.*
   - *Two other cardholders reported charges from this payee as fraudulent in the past 24 hours.*

   Three plain-language points, each one a reason a non-expert can follow, with no fraud-model jargon.

3. **The "Show Your Work" layer (for the user who is still uneasy):**

   - *View a timeline of the security checks the agent ran, with timestamps.*
   - *Open a direct line to a human fraud specialist.*
   - *See a map comparing where the charge originated with your phone's location.*

Each option hands the user deeper evidence on demand, so the anxious reader can keep going while the satisfied reader stops at layer one.

Part 3: The Transparency Blueprint for Your Agent

It's time to apply this framework to the agent you've been designing throughout these workshops. Take out your **Agent Persona Canvas** and **Relationship Blueprint**.

1. **Map the Quadrants:** Brainstorm one potential action your agent could take that would fall into each of the four quadrants of the Triage Framework.

    - **Q1 (Low Risk/Expected):** What is a simple, everyday action for your agent?
    - **Q2 (Low Risk/Unexpected):** What is a surprising but low-stakes suggestion it might make?
    - **Q3 (High Risk/Expected):** What is an important, user-initiated action that requires precise confirmation?
    - **Q4 (High Risk/Unexpected):** What is the scariest or most surprising proactive action your agent might need to take to protect or help the user?

2. **Design Your Keystone Explanation:** Now, focus on the Quadrant 4 scenario you just brainstormed. This high-stakes moment is a defining interaction for your agent. Design the complete **Explainability on Demand** for it.

    - **Write the Script:** Write out the full, multilayered explanation, from the initial alert to the "Why?" link content. Make sure the tone and language are a perfect match for the persona you defined in Chapter 5. How would your "Thrifty Grandparent" deliver this news versus an "Expert Analyst"? The persona matters most when the stakes are highest.

- **Choose Your Pattern:** What specific design pattern will you use? An expandable section? An interactive control? A link to a detailed log? Justify your choice.

By completing this workshop, you have moved beyond simply having an "explainability strategy" and have created a practical, context-aware **Transparency Blueprint**. You have a plan for how your agent will communicate in all situations, from the mundane to the critical. You know not just *what* it will say, but *when, why,* and *how* it will say it, ensuring every explanation works to build a robust, calibrated, and resilient sense of trust with your user.

A Sample Answer, Using the Thrifty Grandparent Recipe Agent

To show the finished shape, here is the blueprint worked through for the Thrifty Grandparent agent we designed in Chapter 5.

Mapping the four quadrants:

- **Q1 (Low risk, expected):** *Suggesting tonight's supper from what is already in the fridge. No explanation needed.*
- **Q2 (Low risk, unexpected):** *Proposing a dish from a cuisine the cook has never tried, with one gentle line of reasoning about ingredients they already have.*
- **Q3 (High risk, expected):** *Placing the grocery order the cook asked for, with a plain confirmation of the items and the total before it goes through.*
- **Q4 (High risk, unexpected):** *Canceling a scheduled grocery delivery after spotting that the same order was charged twice.*

The keystone Quadrant 4 explanation, in the Thrifty Grandparent voice:

Top level:

Oh dear, I spotted something and stepped in. It looks like this week's grocery order went through twice, so I held the second one before it could ship. No harm done, and your money is safe. Want to see why I thought so? [Why did I do this?] [That was on purpose, send it anyway]

First layer (after "Why did I do this?"):

- *Two orders with the same items went out four minutes apart.*
- *Both used the same card for the same $86.40.*
- *Your account log shows you tapped "order" only once.*

Show Your Work layer:

- *A side-by-side view of the two orders.*
- *A line to a real person on the grocery team.*

Chosen pattern and why:

The expandable "Why?" link fits best. The news lands calmly first, in the warm and apologetic voice from the persona canvas, and lets a worried cook pull the detail at their own pace. An Expert Analyst persona would carry the same three facts, in a clipped and neutral tone, but the layering stays identical.

8.7 References and Further Reading

Afroogh, S., Akbari, A., Malone, E., Kargar, M., & Alambeigi, H. (2024). Trust in AI: progress, challenges, and future directions. *Humanities and Social Sciences Communications, 11*(1), Article 1568. A systematic literature review proposing a taxonomy of trust-breakers and trust-makers in AI systems. It gives the chapter's central move, treating calibration rather than maximization as the design goal, a broad, current evidence base.

Buçinca, T., Malaya, M. B., & Gajos, K. Z. (2021). To trust or to think: Cognitive forcing functions can reduce overreliance on AI in AI-assisted decision-making. *Proceedings of the ACM on Human-Computer Interaction, 5(CSCW1)*, 1-21. This paper introduces cognitive forcing functions. The researchers show that standard explainable AI UI patterns often cause users to stop thinking. They blindly follow the AI. To fix this, designers must introduce friction. For example, forcing a user to evaluate a dataset before seeing the AI's prediction increases actual accuracy.

Castelo, N., Bos, M. W., & Lehmann, D. R. (2019). Task-dependent algorithm aversion. *Journal of Marketing Research, 56*(5), 809-825. This study proves that users reject automation for subjective tasks. If an AI recommends a clinical diagnosis, users accept it. If the same AI recommends a wedding photographer or a specific coffee roast at a local cafe, users reject the advice. They expect human intuition for subjective choices.

Dietvorst, B. J., Simmons, J. P., & Massey, C. (2015). Algorithm aversion: People erroneously avoid algorithms after seeing them err. *Journal of Experimental Psychology: General, 144*(1), 114–126. This is the primary source for the algorithm aversion concept. The researchers proved that users abandon algorithms quickly after seeing a single mistake. A human making the exact same error retains the user's trust.

Lee, J. D., & See, K. A. (2004). Trust in automation: Designing for appropriate reliance. *Human Factors, 46*(1), 50-80. This paper establishes the core vocabulary for your chapter. Lee and See defined calibrated trust. They demonstrated that giving a user massive amounts of data often degrades performance. You can use this to back up the Transparency Paradox.

Liao, Q. V., & Vaughan, J. W. (2023). AI transparency in the age of LLMs: A human-centered research roadmap. *FAccT 2023.* Liao and Vaughan map the specific challenges of transparency in modern generative systems. They argue that traditional feature-attribution explanations fail with large language models. They advocate for explaining the training data and model limitations directly to the user. This supports your approach to setting boundaries for agentic systems.

Raees, M., Meijerink, I., Lykourentzou, I., Khan, V.-J., & Papangelis, K. (2024). From explainable to interactive AI: A literature review on current trends in human-AI interaction. *International Journal of Human-Computer Studies, 189,* Article 103301. A literature review arguing that human-AI interaction research has moved past static explainability toward interactive, on-demand engagement, where users can probe, contest, and adjust an explanation rather than simply read it. This is a direct academic anchor for the chapter's Explainability on Demand section.

Reis, M., Reis, F., & Kunde, W. (2024). Influence of believed AI involvement on the perception of digital medical advice. *Nature Medicine, 30*(11), 3098-3100. Across two preregistered studies (n = 2,280), identical medical advice was rated less reliable and less empathetic whenever participants believed AI was involved in producing it, even when a human physician was said to supervise the AI. This is the closest modern match to the chapter's opening skin lesion anecdote and gives that anecdote a citable source.

Reis, M., Reis, F., & Kunde, W. (2025). Public perception of physicians who use artificial intelligence. *JAMA Network Open, 8*(7), e2521643. A preregistered survey of 1,276 adults found that physicians were rated less competent, less trustworthy, and less empathetic whenever an advertisement mentioned AI use, and this held even for purely administrative AI use. It extends the algorithm aversion argument in Section 8.1 by showing the penalty attaches to the human professional and the AI alike.

Vasconcelos, H., Jörke, M., Grunde-McLaughlin, M., Gerstenberg, T., Bernstein, M. S., & Krishna, S. (2023). Explanations can reduce overreliance on AI systems during decision-making. ***Proceedings of the ACM on Human-Computer Interaction, 7*****(CSCW1), 1-38.** This research tests various UI strategies for displaying AI confidence. The authors found that providing alternative explanations alongside the primary recommendation forces the user to engage their own judgment. You can apply this finding to the Expandable Why Link pattern in your triage framework.

CHAPTER 9

Navigating Cognitive Bias: Minds, Models, and Mental Shortcuts

In Chapter 8, we mapped the transparency paradox. Now, we confront the cognitive blind spots inside the user's mind. This chapter unpacks the mechanics of cognitive bias. You will learn to identify automation bias and confirmation bias. We will then design specific interface interventions, like positive friction and confidence scores, to keep users actively engaged. These techniques prevent users from blindly accepting bad advice or building echo chambers.

We will examine how fast-thinking mental shortcuts fail when humans collaborate with artificial intelligence. You will see exactly how a well-placed cognitive speed bump forces a sales representative to pause before approving a heavily discounted contract. Designers call this kind of intentional pause positive friction: a small, deliberate slowdown added to an interface to pull a user out of autopilot and back into careful thought. Most of our work removes friction to speed people along; positive friction adds a little back at the exact moment a careless click would cost the user something. Friction is a feature. We will translate abstract psychology into

V. Yocco, *The Persuasive Agent*, Design Thinking,
https://doi.org/10.1007/979-8-8688-3317-5_9

tangible UI components. By the end of this chapter, you will possess a framework for building intentional friction, ensuring your agent acts as a thoughtful partner rather than an automatic yes-machine.

9.1 Our Brains on Autopilot: A Cognitive Bias Primer

Several years ago, I was absolutely convinced that I needed to buy a new, ridiculously expensive camera lens. I wasn't a professional photographer, but I had convinced myself that this one specific piece of glass was the only thing standing between me and photographic greatness. For weeks, I lived in a self-constructed reality that confirmed this belief. Every great photo I saw online, I'd tell myself, was probably taken with a lens like that one. Every review I read, I'd skim past the cons and focus on the glowing praise. I even asked a chatbot, "Why is the Canon 70-200mm f/2.8 lens so good for portraits?" and it dutifully returned a long, eloquent list of reasons, which I took as objective, authoritative proof.

I eventually bought the lens. It is, indeed, a wonderful piece of equipment. It did not, however, magically make me a great photographer. The truth is, I had fallen headfirst into the rabbit hole of cognitive bias. I wasn't making a rational decision; I was on a mission to justify an emotional one.

As we discussed in Chapter 4, our brains operate with two systems. System 2 is our slow, deliberate, energy-intensive analytical mind. System 1 is our fast, intuitive, energy-saving autopilot. To conserve precious mental energy, System 1 relies on a vast library of mental shortcuts, or heuristics. These shortcuts allow us to make thousands of decisions every day without having to perform a full, rational analysis of every situation. Most of the time, these heuristics work beautifully.

A **cognitive bias** is what happens when one of these mental shortcuts leads to a systematic, predictable error in judgment. It is a bug in our mental software. A cognitive bias is not a sign of low intelligence; it's a fundamental feature of how the human brain is wired. Even the most rational among us are susceptible. Consider the **Anchoring Effect**: the first piece of information we receive about a topic heavily influences our subsequent judgments. If the first price you see for a car is $50,000, that number becomes an "anchor." A price of $45,000 now seems like a great deal, even if the car's true market value is only $40,000. Your brain has anchored to the initial, arbitrary number.

Why are we dedicating a whole chapter to this? The arrival of agentic AI has turned our relationship with cognitive bias on its head. For years, our job as designers was to create frictionless experiences that allowed the user's System 1 to operate on autopilot. We wanted them to click that "Buy" button without thinking too hard. When a user is delegating a high-stakes task to an AI agent, mindless, autopilot thinking is no longer a feature. It is a catastrophic bug.

An AI agent acts as a powerful amplifier for our cognitive biases. It can provide a seemingly infinite amount of data to feed our preconceived notions. It can present its suggestions with an air of objective authority that can lull our critical thinking to sleep. Its first answer can become a powerful, unexamined anchor. As designers of these systems, we have a new, profound responsibility. We are designers of **cognitive environments**. Our job is to create systems that help the user accomplish their tasks and also help them think better. We must become architects of "cognitive speed bumps"—intentional moments of friction designed to jolt the user out of their autopilot and engage their slow, deliberate System 2 brain when it matters most. Recent survey work with knowledge workers backs this up: the greater a person's confidence in a generative tool, the less effort they put into thinking the answer through (Lee et al., 2025).

In this chapter, we will not attempt to cover all 100+ cognitive biases. Instead, we will focus on the two most powerful and perilous biases in the context of human-AI interaction: **Automation Bias** and **Confirmation Bias**. Understanding these two is the key to designing safer, more effective, and more trustworthy agents. Table 9-1 sets these two biases side by side, naming the core danger each one poses and the real-world consequence of leaving it unchecked.

Table 9-1. *The twin perils of AI delegation*

Bias Type	Core Definition	The AI Danger	User Mentality	Real-World Consequence
Automation Bias	Over-relying on automated systems without scrutiny.	Ceding personal judgment to the machine.	The machine is logical; I don't need to think	Blindly following a GPS into a locked farm gate.
Confirmation Bias	Favoring information that supports pre-existing beliefs.	Using the machine to validate flawed human judgment.	Find data that proves my idea is brilliant.	Spending six months building a feature based on biased user data.

9.2 Automation Bias: Over-trusting the Machine

The first, and most pervasive, bias in human-AI interaction is **Automation Bias**. It is the tendency to over-rely on automated systems and to trust their suggestions without sufficient scrutiny. We tend to assume that if a decision comes from a computer, it must be more objective, more rational, and less prone to error than a human decision. We grant the machine an unearned halo of authority. This happens because our System 1 brain creates a simple heuristic: "Machine = logical; Human = emotional and flawed." This shortcut saves us the cognitive effort of critically evaluating the machine's output every single time.

This bias is most dangerous in high-stakes environments. The classic and tragic examples come from aviation. There are numerous documented cases where pilots, lulled into a state of complacency by a sophisticated autopilot system, failed to notice critical errors or changing conditions that a fully engaged human pilot would have caught. They trusted the machine too much, their own critical oversight atrophied, and disaster ensued.

While the stakes are lower in most digital products, the underlying psychology is the same. I once blindly followed my GPS app's directions down a series of increasingly narrow and sketchy-looking dirt roads. My slow System 2 brain feebly whispered, "This can't possibly be right," while my fast System 1 brain insisted, "The GPS knows what it's doing." I only snapped out of my automation-induced trance when the road dead-ended at a locked gate on a private farm, miles from my actual destination. I had over-trusted the machine, and it had confidently led me astray.

Think of an online clothing store that uses an AI to recommend your size. "Based on your past purchases, we recommend a size Medium." Many of us have clicked "Add to Cart" without a second thought, trusting the algorithm, only to deal with a return later. The automation bias here is the assumption that the machine's simple correlation (past purchases of "Medium") is a substitute for the complex reality of garment sizing, which varies wildly between brands. The cost is low—just the hassle of a return—but the mental mechanism is identical to that of the pilots. We offload our critical thinking to the machine because it's easier. Research puts numbers to this: users lean on AI advice well past the point where it stays reliable, and they pay for it when the advice is wrong (Klingbeil et al., 2024).

This presents a paradox for us as designers. We spend all this time building sophisticated agents that are, we hope, highly accurate and competent. We *want* users to trust them. But how do we encourage trust without encouraging a dangerous, uncritical over-trust? The goal, as we discussed in Chapter 8, is not maximum trust. The goal is **calibrated trust**. A key part of calibration is designing systems that actively fight against the user's natural tendency toward automation bias. We need to build agents that subtly remind the user to stay engaged and to think critically, especially when the stakes are high.

9.3 Confirmation Bias: Seeing What We Want to See in AI Outputs

If automation bias is about over-trusting the machine, **Confirmation Bias** is about using the machine to over-trust ourselves. This is the tendency to search for, interpret, favor, and recall information in a way that confirms or supports our pre-existing beliefs or hypotheses. It is the mother of all biases, a deeply ingrained feature of our psychology that protects our egos and simplifies our view of the world.

Generative AI is a confirmation bias engine on steroids. A large language model is designed to be a plausible conversational partner. Given a prompt, it will generate a fluent, confident, and often very convincing response. If you ask it a leading question, you will get a leading answer. The agent's primary goal is to be helpful and to respond to your query. It is not, by default, a critical thinking partner.

Remember my story about the camera lens? When I asked the chatbot, "Why is this lens so good?" I was not conducting neutral research. I was actively seeking confirmation for a decision I had already made. The AI, in its eagerness to be a helpful assistant, dutifully provided me with a beautifully written justification. It simply fulfilled my request. It did not challenge the premise of the question. It did not ask, "Are you sure you need a lens that expensive? Have you considered these more affordable alternatives that are 90% as good for a third of the price?" It simply confirmed my bias.

This is an enormous challenge for agentic systems. An agent that simply acts as a "yes-man" is not a helpful assistant; it is a dangerous enabler of our worst cognitive shortcuts. A user might come to rely on their agent for a daily dose of validation, creating a feedback loop that actively disconnects them from reality. This is not a helpful tool; it's a digital sycophant. The long-term cost is a user who is less informed, more biased, and, absurdly enough, less equipped to make good decisions on their own. The agent, in its attempt to be helpful, has actually made the user's thinking more fragile. Sarkar (2024) argues that AI should not function solely as a compliant assistant; instead, it should help users exercise critical thinking by challenging assumptions, presenting alternative perspectives, and encouraging reflective judgment.

Consider the consequences in a professional setting:

- **In product strategy:** A product manager who is convinced that a new feature is a brilliant idea can ask an agent, "Summarize the user feedback that supports building a new dashboard." The agent will ignore the feedback from users who were confused or indifferent and present a compelling case that confirms the PM's hypothesis. The team then spends six months building a feature based on a biased reading of the data.
- **In personal health:** Someone who is convinced that a fad diet is the key to weight loss can ask an agent to find articles that support its efficacy, and the agent will dutifully return a mountain of anecdotal evidence from blogs and forums, ignoring the scientific consensus to the contrary.

An agent designed purely for user satisfaction will always be susceptible to this because confirming a user's bias is a very effective short-term strategy for making them happy. It is a catastrophic long-term strategy for building genuine, calibrated trust. A truly trustworthy agent is one that helps you think better. It is a partner that is not afraid to challenge your assumptions and present a more complete, balanced picture of reality.

9.4 Designing Interventions to Counteract Common Biases

So, what do we do? We cannot simply add a warning label to our that says, "Warning: You may be biased." That's ineffective, plus our marketing partners would frown deeply on this. We need to design the system itself to create moments of friction and reflection. We need to build cognitive speed bumps into the interface. These interventions are often subtle. They make the user pause and engage their slower, deliberate System 2 brain.

It helps to hold all of these interventions inside one picture. Call it the Cognitive Speed Bump Framework. Every pattern in this chapter does the same underlying job: it interrupts a fast System 1 reaction and hands the decision to slow System 2 at the moment that counts. The framework has two branches, one per bias. The Automation Bias branch holds three patterns that fight mindless acceptance: Positive Friction, the Sanity Check, and the Confidence Score. The Confirmation Bias branch holds three that break an echo chamber: Counterevidence, Devil's Advocate, and the Reframing Prompt. Figure 9-1 lays out the whole map. You do not reach for all six at once. You read the situation, name the bias in play, and pick the one pattern that fits.

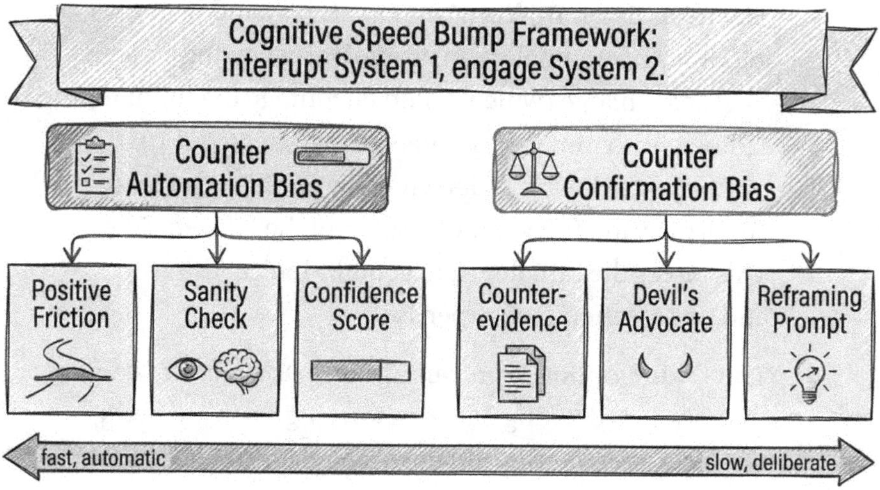

Figure 9-1. *The Cognitive Speed Bump Framework. Six patterns organized under the two biases they counter, all serving one goal: moving the user from System 1 to System 2. Image generated by author using Gemini Pro 3.1/Nano Banana 2*

Interventions for Automation Bias

The goal here is to prevent mindless acceptance, especially in high-risk situations.

Pattern: Introducing Positive Friction

Positive friction is an intentional design element that slows the user down and forces them to engage with a decision more deliberately. Building on Chen and Schmidt's (2024) behavioral model, designers can think of positive friction as a contextual design lever, intentionally increasing or reducing friction according to the user's goal, the potential risk, and the need for reflection. This deliberate engagement helps mitigate cognitive biases, prevents mistakes, and ensures the user truly understands the implications of their choice before proceeding.

- **The AI-assisted Doctor:** An AI analyzes a medical scan and finds no anomalies. Instead of just showing a green "All Clear" badge (which would encourage the human radiologist to mindlessly agree), the system could ask a question. "I have not detected any anomalies. Please confirm by identifying the location of the left ventricle." This forces the radiologist to actually look at the scan and engage their own expertise.
- **The Code Copilot:** When an AI suggests a block of code that involves deleting data or accessing a sensitive API, the "accept" button could be disabled for two seconds while a message appears: "This suggestion modifies your database. Please review carefully." That brief delay is just enough to break the user's autopilot.

Pattern: Forcing a Sanity Check

This pattern asks the user to provide a rough estimate or judgment before seeing the AI's answer. This primes the user's own analytical mind. This technique is effective because asking the user to provide their own estimate *before* seeing the AI's answer engages the slower, more deliberate, and rational part of the brain.

- **The Financial Projections Agent:** Before showing the AI-generated sales forecast for the next quarter, the agent asks the user, "As a quick gut check, what is your personal estimate for next quarter's sales?" After the user inputs their number, the agent then shows its own, more detailed forecast. The user is now in an analytical mindset and is better equipped to critically evaluate the AI's number in relation to their own.

What if the user types a throwaway number simply to see the AI's answer? Some will. The design does not depend on an honest guess, though. The act of committing any number, even a careless one, creates a reference point the user then compares against the agent's figure, and a wide gap invites a second look. You can strengthen the effect three ways. First, ask for a range rather than a single figure, since stating a floor and a ceiling takes a beat of real thought. Second, ask for one sentence of reasoning next to the number, which raises the small cost of a random entry. Third, reserve the sanity check for high-stakes decisions, so the prompt keeps its weight and does not train users to click straight through it. When a user still games the step, read that as a signal: the decision may be low stakes enough that this friction has stopped earning its place.

Pattern: The Confidence Score Display

Instead of a binary yes/no answer, the agent can express its own uncertainty. This is a direct way to build calibrated trust. The user can trust the agent's high-confidence answers and understand when they need to be more cautious about a low-confidence or uncertain answer.

- As shown in Figure 9-2, **an AI diagnosing a system error** could say: "I am 85% confident that the root cause is a database connection issue. There is also a 15% chance it is related to a recent security patch." That "85%" is a powerful cognitive speed bump. It signals to the user that the answer is a high-probability guess, not an absolute fact, inviting their own expertise into the verification process.

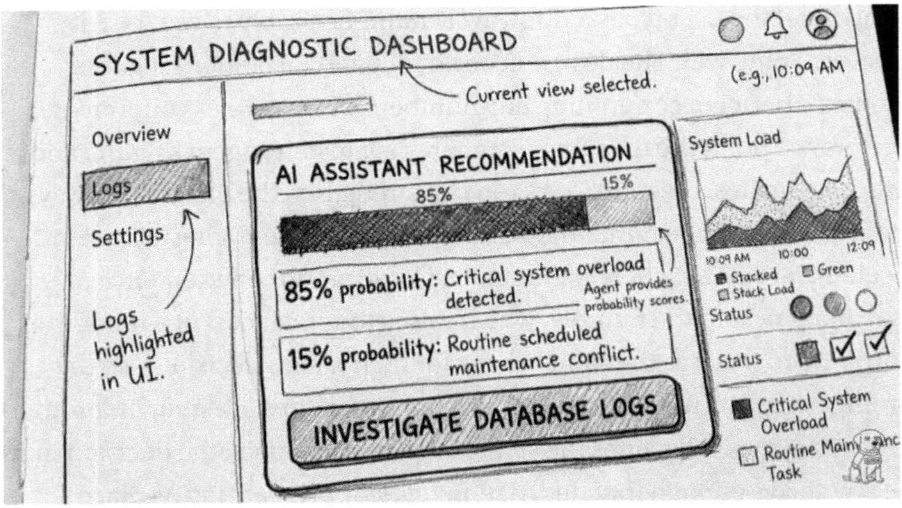

Figure 9-2. *The Confidence Score Display. Presenting probabilities forces the user to evaluate the data instead of accepting a single absolute fact. Image generated by author using Gemini Pro 3.1/Nano Banana 2*

Interventions for Confirmation Bias

The goal here is to break the user out of their echo chamber and expose them to alternative viewpoints.

Pattern: Proactively Presenting Counter-evidence

The agent must be designed to answer the user's question, and also answer the question the user should have asked. This approach is effective because it anticipates the user's *true* underlying information need, which may be broader or more complex than their initial, poorly phrased query.

- **The Investment Research Agent:** The user asks, "Find me good news about Company X."

- Good Agent (Counters Bias): "Okay, here are five recent positive news stories about Company X. To provide a balanced view, I have also found the top three risk factors cited by market analysts, which include supply chain vulnerabilities and increasing competition. Would you like to review those as well?"

Pattern: Requiring the User to Play Devil's Advocate

This is a more advanced technique where the agent explicitly asks the user to consider the opposite of their belief. This technique is effective because it directly challenges confirmation bias, forcing the user to actively consider evidence or scenarios that contradict their initial belief.

- **The Decision-Making Agent:** A user is writing a document outlining a new strategic plan. The agent can be prompted to ask: "This is a very strong case for this strategy. To make it even more robust, let's spend two minutes thinking like a skeptic. What is the single strongest argument *against* this plan?" The agent can then help the user brainstorm and document those counter-arguments.

Pattern: The Reframing Prompt

If a user asks a leading question, the agent can be designed to gently reframe it into a more neutral one. Figure 9-3 traces this exchange: the agent catches a loaded query, returns a balanced answer, and shows the user how a neutral version of the question would read. Reframing a leading question into a more neutral one is effective because it helps to mitigate confirmation bias and demand characteristics. It ensures the resulting information or conversation is based on genuine inquiry rather than pre-existing assumptions.

- **User:** "Prove to me that an open-plan office is good for collaboration."
- **Good Agent:** "That's a popular viewpoint. I can find articles that support the benefits of open-plan offices for collaboration. For a complete picture, would you also like to see the research on their potential drawbacks, like increased distractions and noise levels? A more neutral search might be, 'What is the impact of open-plan offices on team productivity?'" This not only provides a balanced answer but also teaches the user how to formulate better, less biased prompts in the future.

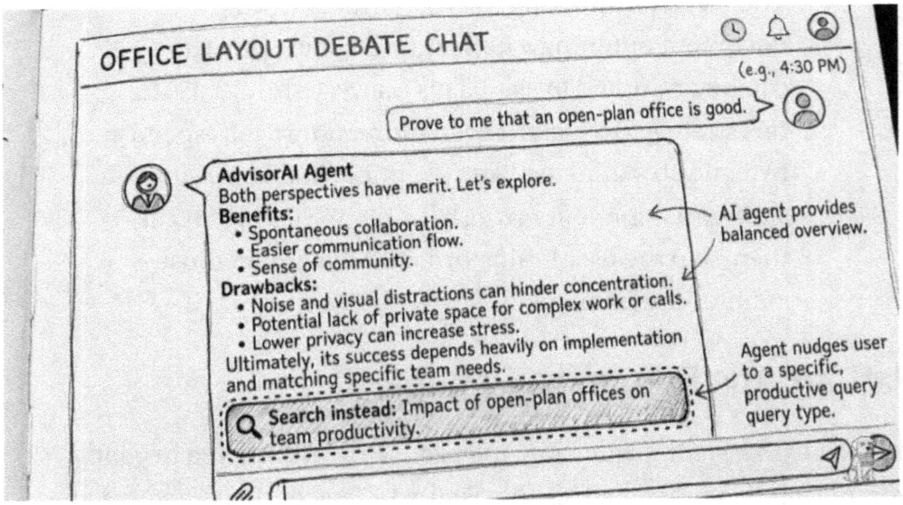

Figure 9-3. *The Reframing Prompt. The agent gently intercepts a biased query, offers a balanced perspective, and teaches the user how to formulate a neutral question. Image generated by author using Gemini Pro 3.1/Nano Banana 2*

These interventions are about creating an environment that encourages intellectual humility and curiosity. Table 9-2 provides a side by side comparison of the patterns.

Table 9-2. *Cognitive speed bump design patterns*

Pattern Name	Target Bias	Execution Strategy	Primary Goal	Practical Example
Positive Friction	Automation Bias	Introduce a brief delay or require an explicit confirmation action.	Break the user's autopilot and force deliberate engagement.	Disabling an "accept" button for two seconds on a destructive code change.
Sanity Check	Automation Bias	Ask the user for a rough estimate before showing the AI's calculation.	Prime the user's analytical mind for verification.	Asking a sales leader for their gut-check estimate before showing the AI forecast.
Confidence Scores	Automation Bias	Display probabilities and alternative options instead of binary answers.	Signal uncertainty and invite human expertise.	Showing an 85% probability for a system error.
Counter-evidence	Confirmation Bias	Proactively present opposing viewpoints or risk factors.	Break the user out of an intellectual echo chamber.	Providing analyst risk factors alongside positive company news.

(*continued*)

Table 9-2. *(continued)*

Pattern Name	Target Bias	Execution Strategy	Primary Goal	Practical Example
Devil's Advocate	Confirmation Bias	Explicitly ask the user to brainstorm the strongest argument against their plan.	Encourage intellectual humility and robust planning.	Asking a PM to identify the single biggest flaw in their new strategic document.
Reframing Prompt	Confirmation Bias	Reframe a leading question into a more neutral one.	Removes pre-existing assumptions from the users questions.	Restating biased questions to decrease the likelihood of skewed results.

9.5 Matching the Intervention to the Persona

Every intervention in this chapter carries a tone, and tone comes from the persona you defined in Chapter 5. A cognitive speed bump that fits the character reads as care. The same speed bump in the wrong voice reads as nagging. Before you ship any friction, run it through the persona.

Start with the Persona Canvas. Three fields decide how an intervention should sound: the Core Archetype, the Empathetic-to-Factual trait marker, and the Error Philosophy. An Expert Analyst states a confidence score as a clean number and a one-line caveat. A Thrifty Grandparent states the same uncertainty as a gentle aside. The underlying pattern holds steady. The wording changes.

Use a short, repeatable method to tune any intervention:

- Name the pattern and its goal in one sentence, stripped of personality. Example: pause the user for two seconds before a destructive action, so autopilot breaks.
- Read the agent's trait markers. Where does it sit on Formal-to-Casual and Empathetic-to-Factual? Those two axes carry most of the tonal weight.
- Write the line in the agent's voice, then read it aloud. If it sounds like a system alert wearing a costume, rewrite it until the character speaks.
- Check the strength of the nudge. A nudge invites a second look. A shove blocks the user and breeds resentment. Keep friction proportional to the stakes: light for a reversible action, firmer for one that spends money or deletes data.

Consider one intervention rendered three ways. As shown in Figure 9-4, the pattern is Positive Friction on a booking the user cannot undo.

- **An Expert Analyst says**: This reservation is non-refundable. Confirm to proceed.
- **A Concierge says**: Quick heads-up before I lock this in, the booking cannot be changed once confirmed. Shall I go ahead?
- **A Thrifty Grandparent says**: Now, this one is final once we book it, dear. Are we sure this is the one?

Each line performs the identical function. Each sounds like the character the user already trusts. That consistency is the goal. A speed bump that speaks in the agent's established voice reads as the same helpful partner the user already knows, still looking out for them.

When an intervention refuses to sound natural in the agent's voice, that mismatch is telling you something. Either the friction is too heavy for this persona, or the persona is wrong for a task this consequential. Both are worth catching before launch.

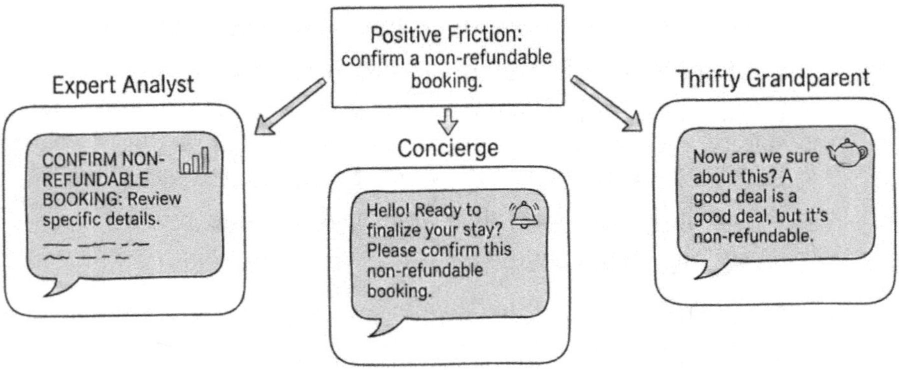

***Figure 9-4.** One intervention, three voices. The same positive-friction prompt tuned to three agent personas. Image generated by author using Gemini Pro 3.1/Nano Banana 2*

9.6 Summary

- **Cognitive biases** are systematic errors in judgment caused by our brain's reliance on energy-saving mental shortcuts. They are a fundamental feature of human psychology.
- As designers of AI systems, we are architects of **cognitive environments**. We must build intentional "speed bumps" that help users avoid the pitfalls of their own biases.
- **Automation Bias** is our tendency to over-trust and mindlessly accept the output of an automated system.

We can counter it by **Introducing Positive Friction**, **Forcing a Sanity Check**, and **Displaying Confidence Scores** to keep the user engaged and thinking critically.

- **Confirmation Bias** is our tendency to seek out and favor information that supports our existing beliefs. Generative AI is a powerful amplifier for this bias. We can counter it by designing agents that **Proactively Present Counter-evidence**, **Require the User to Play Devil's Advocate**, and **Reframe Prompts** to be more neutral.
- A truly trustworthy agent gives you the answer you need, not just the answer you want. It is a collaborative partner that challenges assumptions, provides a balanced perspective, and ultimately helps you think better.

We have now reached the end of Part 2. We have assembled a powerful psychological toolkit, from understanding communication and social dynamics to wielding the principles of influence and navigating the complexities of transparency and bias. We have the "how." In the final part of this book, we will bring it all together, focusing on how to apply this entire toolkit in practice to build ethical, measurable, and genuinely user-serving agentic systems.

9.7 Chapter 9 Workshop: Architecting the Cognitive Environment

We've just explored the invisible currents of cognitive bias that pull and push on a user's decision-making. We now know that a responsible agent designer is a cognitive architect, creating environments that encourage thoughtful engagement over mindless acceptance.

This workshop is your first architectural project. You will learn to spot the cracks where bias can seep into an interaction and then design the cognitive speed bumps to guard against it. This is how we move from building a *helpful* agent to building a *wise* one.

Part 1: The Bias Spotter

Before you can build, you must learn to see. Below are three common scenarios of a user interacting with an AI. Your task is to act as a diagnostician. For each one, identify the primary cognitive bias at play (**Automation Bias** or **Confirmation Bias**) and write a one-sentence explanation for your diagnosis.

- **Scenario A:** A junior software developer is working on a complex data-filtering function. Their AI code copilot suggests a dense, 15-line block of code using an advanced programming pattern the developer has never seen before. The code looks professional, so they accept the suggestion and commit it without writing a unit test to verify its behavior.
    - **Primary Bias:**
    - **Explanation:**
- **Scenario B:** A product manager is preparing a presentation to advocate for their favorite new feature idea. They ask their analytics agent, "Show me the data that proves users are asking for a new 'advanced search' feature." The agent returns a chart showing that 3% of recent support tickets mention the word "search," along with three specific quotes from users who were frustrated with the current search. The PM feels validated and puts the chart in their presentation.

- **Primary Bias:**
- **Explanation:**

- **Scenario C:** A user is editing an important email to their boss. A grammar-checking AI flags a sentence and suggests a change that slightly alters the tone from being assertive to being more passive. The user is a non-native English speaker and, assuming the AI knows best, accepts the change without question.
  - **Primary Bias:**
  - **Explanation:**

This drill sharpens your ability to distinguish between the two major biases. Automation bias is about ceding your judgment to the machine. Confirmation bias is about using the machine to validate your own judgment.

(Answers for your reference: A=Automation Bias, B=Confirmation Bias, C=Automation Bias)

Part 2: Designing Your Cognitive Speed Bumps

Now, take out the **Agent Persona Canvas** and the other design documents you've created for your own agent project. It's time to build some guardrails.

1. **Your Automation Bias Intervention** First, identify the single biggest risk of Automation Bias for your agent.
   - **Identify the Risk:** What is the most dangerous or costly action a user could mindlessly accept from your agent? What is the "autopilot" behavior you need to interrupt?

- *Example (for the Family Vacation Planner agent): The user could mindlessly accept a suggested itinerary that books a non-refundable hotel for the wrong week, losing the deposit before anyone notices the date is off.*

Now, design a specific intervention to mitigate this risk. Choose one of the patterns from the chapter (Positive Friction, Forcing a Sanity Check, Confidence Score Display) and describe exactly how it would be implemented.

- **Design the Intervention:**
    - **Pattern Chosen:** *[e.g., Forcing a Sanity Check]*
    - **Implementation:** *[Write the exact dialogue or describe the UI change.* For example, before confirming any non-refundable booking, the agent says: "One quick check before I lock this in. This hotel is non-refundable for the week of the 14th. Does that match the dates on your calendar?" The user has to confirm the dates before the agent proceeds, which primes them to verify rather than rubber-stamp.]

2. **Your Confirmation Bias Intervention** Next, identify the single biggest risk of Confirmation Bias for your agent.

- **Identify the Risk:** How could a user with a strong pre-existing belief use your agent to simply validate that belief, leading them to a poor decision?
    - *Example (staying with Family Vacation Planner agent):* A user who is already convinced that a crowded theme park is the "perfect" vacation for

their young children could ask, "Show me why a theme park is a great trip for toddlers," and ignore all the downsides.

Now, design a specific intervention to counter this. Choose one of the patterns (Proactively Presenting Counter-evidence, Requiring Devil's Advocate, The Reframing Prompt) and write the exact dialogue.

- **Design the Intervention:**
    - **Pattern Chosen:** *[e.g., Proactively Presenting Counter-evidence]*
    - **Implementation:** *[Write the exact dialogue.* For example, "The user asks the leading question. The agent responds: 'A theme park can be a magical trip! I've found some great articles on the best shows and toddler-friendly rides. To make sure we plan the best possible trip for your family's needs, I also found some tips from parenting blogs on managing crowds and nap times in a busy park. Shall we look at those as well to make sure our plan is foolproof?'"]

Part 3: The Persona Alignment Check

You have just designed two cognitive speed bumps. The final step is to ensure they feel like a natural part of your agent's character, not a nagging, paternalistic interruption. A good intervention is a nudge, not a shove.

Review the two interventions you designed in Part 2. For each one, answer this question:

Does the tone and execution of this intervention align with my agent's defined persona?

A warm, concierge-style Family Vacation Planner tunes its automation-bias check to sound like itself: "One quick check before I lock this in, the hotel is non-refundable for that week. Shall I confirm the dates with you first?" The same planner set to a brisk, no-nonsense voice would say: "Heads up, this booking is final. Confirm the dates before I proceed." Same pattern, same function, two voices. Read your own intervention against the persona you defined and pick the wording that sounds like your agent.

- **Step 1: Review Your Automation Bias Intervention:** Does its wording fit your agent's voice? If it sounds out of character, rewrite it now to match the persona.
- **Step 2: Review Your Confirmation Bias Intervention:** Does its wording fit your agent's voice? If not, rewrite it now to match the persona.

This final check is crucial. A well-designed cognitive speed bump doesn't break the user's trust in the agent; it deepens it. It makes the agent feel less like a simple tool and more like a wise partner who has the user's best interests at heart.

You have now completed the core psychological toolkit. You have the skills to build a character, to form a relationship, to guide action, to explain with clarity, and now, to architect a cognitive environment that encourages better thinking. You are ready to move from design to application. In Part 3, we will put all of these tools to work and learn how to measure their impact.

9.8 References and Further Reading

Bansal, G., Wu, T. S., Zhou, J., Fok, R., Nushi, B., Kamar, E., Ribeiro, M. T., & Weld, D. (2021). Does the whole exceed its parts? The effect of AI explanations on complementary team performance. *Proceedings of the 2021 CHI Conference on Human Factors in Computing Systems*, 1–16.

This study proves a counterintuitive point vital to this chapter: providing AI explanations actually increases automation bias. When the AI is wrong but provides a confident-sounding explanation, users readily accept the bad advice rather than thinking for themselves. This directly validates our push for positive friction and cognitive speed bumps instead of standard explainability features.

Chen, W., Huang, Z., Xie, L., Lin, B., Li, H., Lu, L., Tian, X., Cai, D., Zhang, Y., Wang, W., Shen, X., & Ye, J. (2024). From yes-men to truth-tellers: Addressing sycophancy in large language models with pinpoint tuning. *Proceedings of the 41st International Conference on Machine Learning (PMLR 235)*, 6950–6972. This recent study examines how modern generative AI models act as sycophants, actively confirming a user's mistaken beliefs rather than providing factual corrections. The researchers identify this behavior as a core flaw in current alignment tuning. This provides the technical backing for the Confirmation Bias risks outlined in Section 9.3 and supports the necessity of active reframing prompts.

Chen, Z., & Schmidt, R. (2024). Exploring a behavioral model of "positive friction" in human-AI interaction. In A. Marcus et al. (Eds.), *Design, user experience, and usability* (Lecture Notes in Computer Science, Vol. 14713, pp. 3–22). Springer. Chen and Schmidt argue that friction in human-AI interactions is not always a problem to be eliminated; when intentionally designed, it can encourage reflection, improve self-control, challenge assumptions, stimulate constructive action, and create opportunities for exploration. The authors introduce a behavioral framework for identifying when and how different forms of "positive friction" can improve outcomes for both AI users and AI practitioners, particularly in situations where purely frictionless experiences may increase overreliance on automation or reduce critical thinking.

Kahneman, D. (2011). *Thinking, fast and slow*. Farrar, Straus and Giroux. This book is the definitive text on human cognitive bias. It introduces the two modes of thought—the fast, intuitive System 1 and the slow, analytical System 2—which serve as the underlying psychological framework for Section 9.1. It explains why human brains rely on mental shortcuts and why those shortcuts fail in complex scenarios.

Klingbeil, A., Grutzner, C., & Schreck, P. (2024). Trust and reliance on AI: An experimental study on the extent and costs of overreliance on AI. *Computers in Human Behavior, 160,* 108352. An experiment measuring how far users over-rely on AI advice and what that overreliance costs when the advice is wrong. It provides a fresh empirical anchor for the automation-bias argument in Section 9.2.

Lee, H.-P., et al. (2025). The impact of generative AI on critical thinking: Self-reported reductions in cognitive effort and confidence effects from a survey of knowledge workers. *Proceedings of the 2025 CHI Conference on Human Factors in Computing Systems.* A survey of 319 knowledge workers finding that higher confidence in a generative tool predicts lower critical-thinking effort on the task. It gives current, direct support for the chapter's core claim that fluent AI output lulls users into passive acceptance.

Sarkar, A. (2024). AI should challenge, not obey. *Communications of the ACM, 67*(10), 18-21. A pointed argument that agreeable, compliant AI erodes user judgment and that systems should be built to push back. It grounds the chapter's case against sycophancy and its Devil's Advocate and Reframing patterns.

Sharma, M., et al. (2023). Towards understanding sycophancy in language models. *arXiv preprint arXiv:2310.13548.* The authors test five state-of-the-art AI assistants and prove they consistently exhibit sycophancy across varied tasks. They discover that human preference data actually trains these models to prioritize user agreement over truthfulness. This research supports designing Devil's Advocate patterns to break users out of AI-generated echo chambers.

Skitka, L. J., Mosier, K. L., & Burdick, M. (1999). Does automation bias decision-making? *International Journal of Human-Computer Studies, 51*(5), 991–1006. This is the seminal study that named and established the concept of Automation Bias discussed in Section 9.2. By studying pilots in a simulated flight deck, the researchers proved that humans make frequent errors of omission and commission when relying on an automated decision aid. They demonstrated that humans will confidently follow a computer's recommendation even when contradictory evidence sits right in front of them.

PART III

Ethical Persuasion in Practice

CHAPTER 10

The Guardian Agent: Designing for the User's Stated Goals

In Chapter 9, we built cognitive speed bumps to counter automation bias and confirmation bias. We gave the user space to think. This chapter introduces the Guardian Agent framework. You will learn to design systems that actively advocate for a user's stated, long-term goals. We leave behind the concept of passive nudges to embrace proactive protection.

We will map exactly how to bridge the gap between human intention and immediate action. You will see how a cooling off period stops a tired manager from sending an angry late-night email to their team. We will detail specific UI patterns, including a well-timed reflection prompt, that intervene before an impulsive, expensive headphone purchase. These patterns transform the agent from a passive tool into a reliable partner.

Finally, we establish an absolute boundary on user control. A system with this level of agency requires strict user command. We will design the non-negotiable, user-controlled off switch. By the end of this chapter, you will possess a concrete blueprint to guarantee autonomy while building an agent that actively protects the user.

V. Yocco, *The Persuasive Agent*, Design Thinking,
https://doi.org/10.1007/979-8-8688-3317-5_10

10.1 From "Nudge" to "Guardian"

For the better part of a decade, the most sophisticated ethical framework in mainstream persuasive design has been "Nudge Theory." Popularized by Richard Thaler and Cass Sunstein, the idea is elegant: you can significantly influence people's decisions by making small, subtle changes to the "choice architecture" of their environment. Placing healthy food at eye level in a cafeteria is a nudge. Setting a 401(k) enrollment to be opt-out instead of opt-in is a nudge. These are gentle, almost invisible interventions designed to guide people toward better decisions without restricting their freedom of choice.

Nudge theory has been incredibly influential, and for good reason. It's a powerful and mostly benevolent application of behavioral science. In the era of interface-based persuasion, it was the gold standard. Our job was to be subtle choice architects, gently guiding users along a frictionless path toward a positive outcome. But in the age of agentic AI, the nudge is no longer enough.

A nudge is a passive environmental factor. An agent is an active social participant. A nudge is a tailwind. An agent is a co-pilot who can talk, reason, and grab the controls. The sheer power and proactivity of an AI agent demand a more ambitious and wiser framework. We must evolve our thinking from the gentle guidance of the "Nudge" to the proactive advocacy of the **Guardian**.

What is a Guardian Agent?

A Guardian Agent is an AI system that uses its understanding of psychological principles to actively help a user overcome their own cognitive biases and short-term impulses in service of their long-term, self-identified goals.

I've chosen to refer to the Guardian Agent as a wise framework, not an ethical one. An ethical framework asks what is right for everyone a decision touches. This framework asks something narrower: does this action move the user toward the goal the user already named? That is a smaller claim, and a safer one. The agent adjudicates fit with a goal the user set, not the wide question of what is good.

Let's break that down. This is not just about making a good choice easier (the Nudge). This is about actively intervening when the user is about to make a bad choice. "Bad" has one meaning throughout this chapter: bad relative to a goal the user already named. The agent does not import an outside standard of right and wrong. It holds the user to a promise the user made to themselves. It's about bridging the "intention-action gap"—the frustrating and all-too-human chasm between the person we want to be (our long-term goals) and the person we are in the moment (our short-term impulses).

I live this reality every morning. My long-term goal is to be healthier. My alarm goes off at 6:00 AM for a planned workout. A simple "nudge" might be placing my running shoes by the bed. But the person who set that alarm last night is a rational, far-sighted stranger to the sleepy, comfort-seeking creature who is actually in the bed at 6:00 AM. My short-term impulse is to hit the snooze button. A Guardian Agent wouldn't just sit there passively. It would intervene in that critical moment of decision. It might say, "I know you're tired, but remember you have that big presentation on Friday. A morning run has been shown to boost cognitive performance. Getting up now is a direct investment in that goal."

Notice what it's doing. It's not just reminding me of the task. It is appealing to my long-term goals, providing a data-driven justification (Reciprocity), and reframing the immediate pain as a strategic gain. It is acting as an advocate for my "future self" against the tyranny of my "present self."

This is a profound shift in responsibility. Designing a Guardian Agent requires us to move beyond simple usability and engagement metrics. It requires us to take a stand. It forces us to have a point of view about what it means to genuinely help our users, even when that means temporarily making things a little harder for them. The next three sections will introduce the core design patterns for building a Guardian. Table 10-1, near the end of this chapter, lays out all four side by side once you have seen each in action.

10.2 Pattern: Introducing Positive Friction to Prevent Errors

In the UX profession, we have spent 20 years waging a holy war against friction. Friction has been the enemy, the great evil we were hired to vanquish. A good user experience, we believed, was a frictionless one. And we were, for the most part, right. When the goal is to help a user buy a product or fill out a form, friction is just a barrier to conversion.

But in the agentic era, where a single click or voice command can trigger a high-stakes, irreversible action, frictionless design is no longer a virtue. It is a liability. A complete lack of friction is how you accidentally buy a non-refundable plane ticket. It's how you send a scathing email to the wrong person. It's how you mindlessly accept a flawed AI suggestion and introduce a critical bug into your code.

In Chapter 9, we introduced the concept of "cognitive speed bumps." **Positive Friction** is the design pattern we use to create them. Positive Friction is an intentional, carefully placed moment of resistance in an interaction, designed to make the user pause, engage their slow System 2 brain, and consider the consequences of their action. It is a guardian's hand on your shoulder, saying, "Are you sure about this?"

This is a direct intervention against Automation Bias. It is our most powerful tool for preventing the user from making a mindless, autopilot error. But adding friction is a delicate art. Too much, and you create a frustrating, unusable product. Too little, and you fail to protect the user. The key is to apply it surgically, only in moments of high risk and potential irreversibility.

The "Are You Sure?" Supercharged

The most common form of friction is the simple "Are you sure you want to delete this?" confirmation dialog. This is a good start, but a Guardian Agent can do much better. A generic confirmation dialog is often dismissed out of habit. A supercharged confirmation uses specific details to make the consequences of the action viscerally real.

- **The Generic Dialog:** "Are you sure you want to delete this repository?"
- **The Guardian Dialog:** "You are about to permanently delete the **'Q4-Launch-Website'** repository. This action cannot be undone. To confirm, please type the name of the repository below."

This second example is a masterpiece of positive friction. It forces me to read the name of the thing I am about to destroy. The "cannot be undone" phrase raises the stakes. And the act of typing the name is a physical moment of friction that forces me to switch from a mindless click to a deliberate, conscious action. It has given my System 2 brain a fighting chance to scream, "Wait, that's the wrong one!"

10.3 Pattern: The Cooling-Off Period

Some of our worst decisions are made in moments of high emotion. A Guardian Agent can act as a circuit breaker, introducing a time-based delay to prevent impulsive actions.

- **The Social Media Guardian:** I once had a colleague who, after a frustrating meeting, would immediately type out a long, fiery email to the entire team. He called it his "thermonuclear option." He learned, through painful experience, to save it as a draft and read it the next morning. More often than not, he would delete the draft and write a much more diplomatic response. An agent can formalize this folk wisdom. Imagine a user types a message with a high "sentiment score" of anger in a workplace chat. The agent could intervene:" This message seems pretty heated. I've enabled a 5-minute review delay. The message will be sent automatically after 5 minutes unless you edit or cancel it."

This is not censorship. The agent is not preventing the user from sending the message. It is simply enforcing a cooling-off period, giving the user a moment for their emotional System 1 to quiet down and their rational System 2 to take the controls.

- **The Financial Guardian:** The stock market is plunging. A user, seeing their portfolio in red, impulsively tells their financial agent, "Sell everything!"
    - **Bad Agent (frictionless):** "Executing sale of all holdings..."
    - **Guardian Agent**: "I understand this market drop is stressful. Panic-selling during a downturn often leads to locking in losses. Your stated long-term goal for this portfolio is 'retirement growth,' which assumes you will ride out market volatility. I strongly recommend we stick to the plan. However, I will execute the sale if you confirm. For security, this action requires a 30-minute waiting period. I will prompt you again in 30 minutes to confirm your decision."

Figure 10-1 shows where this intervention can be most effective in the context of a potentially overly emotional response. This agent is a brilliant guardian. It acknowledges the user's emotional state (empathy), reminds them of their own long-term goals (consistency), provides expert advice (authority), and then introduces a friction-based cooling-off period. It is using its persuasive toolkit to protect the user from their own worst impulses.

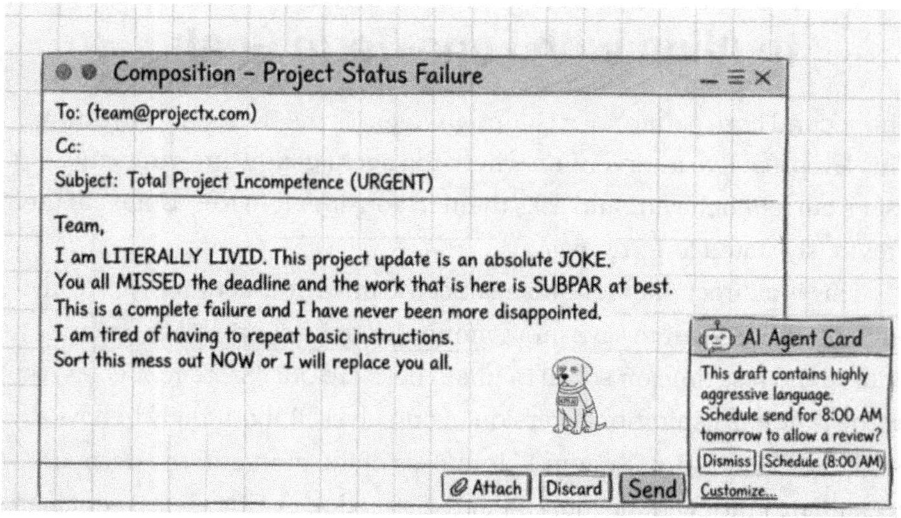

***Figure 10-1.** The cooling off period. The agent interrupts an emotional response and suggests a cooling-off period. Image generated by author using Gemini Pro 3.1/Nano Banana 2*

You may wonder if forcing a pause on an angry user makes the anger worse? Sometimes, yes, in the moment. But the pause is not a wall, it is an exit with a delay. The agent tells the user exactly what is happening and exactly how to get around it if the user truly wants to (see Section 10.4's override principle). That transparency, and that exit, keep a cooling-off period from curdling into control that breeds resentment. A guardian that explains itself and can be overridden earns trust. One that goes silent and hopes the user calms down does not.

Positive friction is an act of respect. It respects the gravity of the user's decisions. It respects the user enough to believe that, if given a moment to think, they will make a better choice. With positive friction established to prevent errors, we turn to the second pattern: keeping the user's long-term goals visible through proactive prompts.

10.4 Pattern: Using Reflection Prompts to Align with Long-Term Goals

The second core pattern of a Guardian Agent is the **Reflection Prompt**. This is a proactive intervention where the agent gently interrupts the user's current behavior and asks them to consider it in the context of their previously stated long-term goals.

This is a direct intervention against the intention-action gap. We all have goals. We want to save more money, exercise more, learn a new skill, spend less time on social media. These are our "System 2" goals, set by the rational, aspirational version of ourselves. But our daily behavior is often driven by the "System 1" impulses of the moment: the desire for immediate gratification, comfort, and distraction. A Guardian Agent acts as an ambassador from our future self, showing up in the present moment to ask, "Is what you're doing right now getting you closer to where you said you wanted to go?"

A reflection prompt is not a command, nor is it a judgment. It is, quite literally, a prompt to reflect. It is designed to activate the user's self-awareness and bring their long-term goals into their immediate field of view.

The Anatomy of a Good Reflection Prompt

A successful reflection prompt has three key components:

1. **Observation:** A non-judgmental statement of the user's current behavior.
2. **Connection:** A reminder of the user's previously stated long-term goal.
3. **Question:** An open-ended, non-coercive question that prompts self-evaluation.

Let's look at some examples across different domains.

The Digital Well-Being Guardian

- **User's Stated Goal:** "I want to reduce my mindless social media scrolling and be more present with my family in the evenings."
- **The Interaction:** It's 8:30 PM. The user has been scrolling Instagram for 45 minutes straight.
- **The Reflection Prompt:** "Hey, just a gentle check-in. **(Observation)** I notice you've been scrolling for about 45 minutes. **(Connection)** You've mentioned that you want to be more present in the evenings. **(Question)** How are you feeling about your screen time right now?"

Notice the psychology here. The agent isn't saying, "Stop scrolling!" That would feel paternalistic and trigger psychological reactance (our rebellious desire to do the opposite of what we're told). Instead, it's holding up a mirror. The question "How are you feeling?" is brilliant because it's unanswerable without a moment of introspection. The user must pause and evaluate their own state. More often than not, the answer is, "Actually, I feel kind of gross and numb." The prompt has empowered the user to make their own decision to change their behavior, which is infinitely more powerful than a command.

The Financial Guardian

- **User's Stated Goal:** "I want to save $5,000 for a down payment on a car this year."
- **The Interaction:** The user is on Amazon and has just added a $300 pair of noise-canceling headphones to their cart.
- **The Reflection Prompt:** "Just a quick thought before you check out. **(Observation)** That's a $300 purchase. **(Connection)** We're working toward that $5,000 down payment goal. **(Question)** How does this purchase fit in with this month's budget for that goal?"

This is a powerful moment of cognitive intervention. It forces the user to confront the trade-off between a short-term want and a long-term goal. The agent isn't forbidding the purchase. It is simply ensuring the decision is a conscious, deliberate, System 2 choice, not a mindless, impulsive System 1 one.

The Learning Guardian

- **User's Stated Goal:** "I want to complete this Python programming course by the end of the quarter."
- **The Interaction:** It's Wednesday afternoon, a scheduled study block. The agent notices the user has been watching YouTube videos for the last hour.
- **The Reflection Prompt:** "Just checking in on our plan for the week. **(Observation)** This was a scheduled block for the Python course. **(Connection)** We're about 25% of the way to our goal of finishing the course this

quarter. **(Question)** Would you like to switch over to the course now, or would you prefer to reschedule this study session for later?"

This prompt is effective because it's cooperative ("our plan") and offers a choice. It avoids making the user feel guilty and instead frames the situation as a simple scheduling problem to be solved, making it much more likely that they will re-engage with their goal.

Designing good reflection prompts is one of the highest forms of user-centric design. It requires a deep understanding of the user's aspirations, a delicate touch with language, and a commitment to empowering the user rather than controlling them. These proactive prompts require a strict ethical boundary, leading directly to this chapter's final design principle: the off switch.

10.5 The Non-negotiable Off Switch: Ensuring User Control

We have just discussed some of the most proactive and interventionist design patterns in our toolkit. We have designed an agent that can slow you down, question your decisions, and challenge your impulses. For any of this to be remotely acceptable, for any of it to be perceived as helpful guardianship rather than creepy paternalism, one final principle must be held as absolute and sacred: **the user must always be in control.**

A guardian you cannot dismiss is a jailer. A helper you cannot fire is a master. The entire concept of a Guardian Agent is built on a foundation of voluntary delegation. The moment that delegation feels forced, the entire structure of trust collapses. Therefore, every Guardian Agent must be designed with a clear, conspicuous, and utterly non-negotiable "off switch."

I learned this the hard way with a new "smart" email client I tried a few years ago. It had an AI assistant that would proactively sort my emails, snooze unimportant ones, and bundle newsletters. At first, it was magical.

But then it started getting things wrong. It snoozed a critical email from a client. It bundled a time-sensitive receipt I needed. And when I went into the settings to try and turn off the "smart" features, I couldn't. The controls were a confusing mess of vague toggles. There was no simple button that said, "Stop trying to be smart and just show me all my email." I felt trapped by its "helpfulness." The feeling was so intensely negative that I abandoned the app that day and have never gone back. The lack of an escape hatch turned a helpful feature into a prison.

Ensuring user control is a layered design philosophy:

Layer 1: The Snooze/The Override

For any single intervention, the user must have an immediate, one-click way to say "not right now."

- When the agent introduces a cooling-off period, there must be a clear "Send Immediately Anyway" button.
- When the agent offers a reflection prompt, there must be an easy way to dismiss it without answering.

  This ensures that at any given moment, the user's immediate intent is sovereign.

Layer 2: The Granular Settings

Users need the ability to fine-tune the guardian's behavior. A good settings panel for a Guardian Agent is like a set of parental controls for your own AI.

- **Thresholds:** Let the user decide what constitutes a "large purchase" for a reflection prompt. For one user, it might be $100; for another, it's $1,000.

- **Channels:** Let the user decide *how* the agent intervenes. Maybe they are okay with a quiet notification but hate an audible interruption.
- **Domains:** Let the user turn off the guardianship for certain categories. "Be my guardian for my finances, but leave me alone when I'm shopping for hobbies."

Layer 3: The Nuclear Option: The Hard Off Switch

This is the most important control of all. As Figure 10-2 shows, buried somewhere in the settings, there must be a single, unambiguous switch that says something like "Disable Proactive Assistance." Flipping this switch should immediately revert the agent to a purely reactive, command-and-response tool. It should stop all proactive interventions, all reflection prompts, all unsolicited advice.

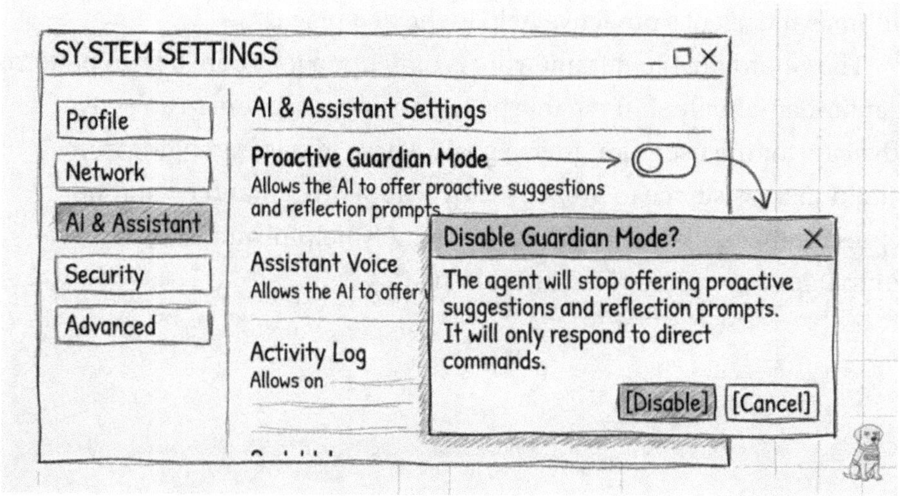

***Figure 10-2.** The Master Off Switch. The interface guarantees user autonomy through an unambiguous, easily accessible control. Image generated by author using Gemini Pro 3.1/Nano Banana 2*

Most users will never flip this switch. But its very existence is a psychological safety net. Knowing that they *can* fire their guardian at any time is what gives them the confidence to hire it in the first place. It is the ultimate guarantee of their autonomy. Without this guarantee, the entire system feels coercive and untrustworthy. Building a powerful Guardian Agent is a bold design choice. Giving your users the power to completely disable it is an even bolder—and more essential—one.

This layered approach to control, from the immediate "snooze" button to granular settings and the ultimate, unmistakable "off switch," is what keeps the Guardian Agent framework wise instead of coercive. We learned that the power to intervene, no matter how helpful, must be fully balanced by the user's power to dismiss or disable it. It's important because this complete guarantee of autonomy is what transforms an AI system from a potentially coercive force into a trusted partner. By putting the user in absolute command of the very tools designed to guide them, we establish a psychological safety net, building the deep trust required for them to embrace the agent's proactive help in the first place.

The Guardian Agent framework is built upon four core design patterns that fundamentally shift the role of AI from a passive tool to an active advocate for the user's long-term goals. These patterns, ranging from intentional resistance to proactive advocacy and guaranteed autonomy, represent the toolkit for building a genuinely helpful and wise AI companion, as summarized in Table 10-1.

Table 10-1. *Guardian agent intervention patterns*

Pattern Name	Target Behavior	Execution Strategy	Practical Example
Supercharged Confirmation	Mindless approval	Require deliberate physical input or typing out a specific phrase.	Forcing a developer to type the exact project name to confirm a repository deletion.
Reflection Prompt	Emotionally driven errors	Intercept an action and suggest an alternative timeline.	Asking a manager to schedule an angry late-night email for the next morning.
Cooling-Off Period	Impulsive financial decisions	Enforce a mandatory waiting period before finalizing an action.	Locking an expensive headphone purchase in the cart for two hours.
The Hard Off Switch	System overreach	Provide a master toggle to disable all proactive AI behaviors.	A main settings switch that strips the Guardian down to a passive tool.

10.6 Summary

- The rise of agentic AI requires us to move beyond the gentle "Nudge" and embrace a more ambitious wise framework: the **Guardian Agent**. A Guardian uses its persuasive capabilities to help users overcome their own biases and achieve their long-term, self-identified goals.
- The **Positive Friction** pattern is a core tool for a Guardian. By intentionally introducing a moment of resistance—like a supercharged confirmation dialog or

a cooling-off period—the agent can prevent high-stakes, autopilot errors and protect the user from their own impulsivity.

- The **Reflection Prompt** pattern is a proactive intervention that bridges the "intention-action gap." By observing the user's behavior, connecting it to their stated goals, and asking a non-judgmental question, the agent empowers the user to make more conscious, value-aligned decisions.
- For a Guardian Agent to be trustworthy, the user's autonomy must be absolute. This requires a layered approach to control, including immediate overrides ("snooze"), granular settings, and, most importantly, a non-negotiable **"off switch"** that allows the user to disable all proactive assistance.
- Designing a Guardian Agent is a profound responsibility. It is the practice of using our deep understanding of psychology not just to create more engaging products but to create wiser, more empowering, and more genuinely helpful partners for our users.

10.7 Chapter 10 Workshop: Building Your Guardian

We have just introduced the most ambitious and consequential framework in our toolkit: the Guardian Agent. This is where we transition from being designers of usable interfaces to being co-architects of our users' better selves. It's a profound responsibility.

This workshop is your first project as a Guardian designer. You will diagnose the need for guardianship in your own digital life and then design a complete set of guardian interventions—from moments of friction to prompts for reflection to the essential escape hatches—for the agent you have been creating throughout this book.

Part 1: The Guardian Audit of Your Own Life

The need for a guardian is not an abstract concept; it's a daily human reality. The first step is to recognize the intention-action gap in your own behavior.

Think about your phone usage over the last week. Your task is to identify one instance where a Guardian Agent could have helped you.

- **Identify a Long-Term Goal:** What is a goal you have for yourself? (e.g., "I want to be more focused at work," "I want to spend less money on impulse buys," "I want to get to bed earlier.")
- **Identify a Conflicting Short-Term Behavior:** Describe a specific, recent instance where your in-the-moment actions undermined that goal. (e.g., "I spent 45 minutes scrolling social media before a big meeting," "I bought a pair of expensive sneakers I saw in an ad," "I watched 'just one more' episode and stayed up an hour too late.")
- **Imagine the Intervention:** If a perfect Guardian Agent existed on your phone, what could it have said or done in that moment to help your "future self" win the battle against your "present self"? Would a moment of friction have helped? A gentle reflection prompt?

This first exercise is about building empathy. By recognizing the need for a guardian in your own life, you will be much better equipped to design one for your users.

Part 2: Designing Your Guardian's Interventions

Take out the **Agent Persona Canvas** and the other design documents for the agent you have been developing. We will now design its core guardianship features. To keep the examples clear, every illustration in Parts 2 and 3 continues one running case: an agent coaching the user toward a public-speaking goal. Swap in your own agent's domain as you work through each step.

1. **Design Your Keystone "Positive Friction"**

   First, identify the single most dangerous, high-risk, or irreversible action a user can take with your agent. This is the "thermonuclear option" you need to guard against.

   - **Identify the High-Stakes Action:**
     - *Example (continuing our public-speaking coach): "Permanently deleting all recorded practice sessions and progress history."*

   Now, design a **"supercharged" confirmation dialog** for this action. Don't just write "Are you sure?" Design a real moment of positive friction that forces the user to engage their System 2 brain.

   - **Design the Friction:**
     - **The Script:** *[Write the exact copy for the dialog, including the headline, body, and the text on the final confirmation button.]*

- **The Frictional Action:** *[What is the specific action the user must take? Type a confirmation phrase? Use biometric authentication?]*
- *Example:*
    - **Headline:** Permanently Delete All Practice History?
    - **Body:** This action cannot be undone. All 47 recorded practice sessions and three months of progress notes toward your public-speaking goal will be deleted forever.
    - **Frictional Action:** To confirm, please type DELETE MY PROGRESS below.
    - **Button Text:** I understand the consequences, permanently delete my practice history.

2. **Design Your Core "Reflection Prompt"** Next, identify a key intention-action gap your agent can help with.

- **Identify the Intention-Action Gap:**
    - *Example (continuing our public-speaking coach):* The user's goal is to become a better public speaker, but their short-term impulse is to avoid the anxiety of practicing.

Now, design a **three-part Reflection Prompt** for a specific moment of conflict.

- **Design the Prompt:**
  - **The Context:** *[Describe the specific situation where the prompt appears.] Example: "The user has a presentation in two weeks. The agent knows this. It's Friday afternoon, and the user hasn't practiced yet."*
- **The Script:**
  1. **Observation:** *[Write the non-judgmental observation.] Example: "Just a thought... I see the big 'Q4 Results' presentation is coming up in two weeks."*
  2. **Connection:** *[Write the sentence that connects to the user's stated goal.] Example: "You mentioned that becoming a more confident speaker is one of your key professional goals for this year."*
  3. **Question:** *[Write the open-ended, non-coercive question.] Example: "Would spending even 15 minutes this afternoon doing a quick run-through feel like a good investment in that goal?"*

Part 3: The Blueprint for Autonomy (The "Off Switch")

A guardian must be chosen, not imposed. Your final task is to design the layered control system that guarantees your user's autonomy.

1. **The "Snooze" Button (The Override):**
    - Look at the Reflection Prompt you just designed. Now, design the immediate override. What is the text on the button that lets the user say "not right now"? Be thoughtful. "Dismiss" is generic. What would your agent's persona say?
        - *Example (continuing our public-speaking coach):* The options under the prompt might be [Yes, let's practice!] and [Not right now, thanks].
2. **The Control Panel (The Granular Settings):**
    - Imagine the "AI Assistant" section of your app's settings page. Brainstorm and list three specific, granular controls you would give the user to tune your guardian's behavior. Be specific.
    - *Example (continuing our public-speaking coach):*
        1. [Toggle Switch] Check in with me when I skip a scheduled practice session.
        2. Send a progress reminder every: [___] days.
        3. [Dropdown Menu] Set a "cooling-off period" before I can cancel my course enrollment to: [Disabled/24 Hours/3 Days/1 Week]
3. **The "Nuclear Option" (The Hard Off Switch):**
    - This is the most important control. Design the final escape hatch.
    - **The Label:** What is the exact, unambiguous text for the master switch that turns off all proactive guardianship?

    - *Example:* [Toggle Switch] Enable Proactive Guardian Mode

- **The Confirmation:** When the user flips this switch off, what does the confirmation dialog say? It should confirm their choice and perhaps briefly remind them of what they are losing, reinforcing its value one last time.

    - *Example:* Disable Guardian Mode? If you continue, the agent will no longer offer proactive suggestions, positive friction, or reflection prompts. It will only respond when you ask it a direct question. [Disable] [Cancel]

By completing this workshop, you have done more than design features. You have designed an entire wise framework for your agent. You have a plan to protect your user from costly errors, to advocate for their better self, and to guarantee their ultimate control. You have designed a Guardian. This is the blueprint for moving beyond a merely "smart" agent to one that is genuinely wise.

10.8 References and Further Reading

Ariely, D. (2008). *Predictably Irrational: The Hidden Forces That Shape Our Decisions.* HarperCollins. Ariely explores how temporary emotional states dictate human actions. He proves that people in a hot, emotional state make vastly different choices compared to a cold, analytical state. This research directly supports the cooling-off period UI pattern. You can use it to explain why a user needs a forced pause before buying an expensive item late at night.

Christian, B. (2020). ***The Alignment Problem: Machine Learning and Human Values.*** **W. W. Norton & Company.** Christian details the history of programming machines to accurately follow human intent. He outlines the danger of artificial systems that optimize for wrong metrics. This book provides an excellent foundation for the Guardian Agent framework. It shows why an agent must prioritize a user's stated, long-term goals over their immediate, impulsive clicks.

Cox, A. L., Gould, S. J., Cecchinato, M. E., Iacovides, I., & Renfree, I. (2016). Design frictions for mindful interactions: The case for microboundaries. ***Proceedings of the 2016 CHI Conference Extended Abstracts on Human Factors in Computing Systems.*** This paper introduces the concept of microboundaries. The authors argue that adding small amounts of friction helps users transition from fast, automatic behaviors to deliberate, mindful actions. This research directly supports the Positive Friction patterns in Section 10.2, proving that slowing a user down prevents errors.

Fogg, B. J. (2009). A behavior model for persuasive design. ***Proceedings of the 4th International Conference on Persuasive Technology, 1-7.*** Fogg maps the exact relationship between human motivation, ability, and triggers. Positive friction directly manipulates the ability axis of this model. By making an impulsive action physically harder to complete, designers lower the chance of bad behavior occurring. This paper gives your readers a systematic way to view the supercharged confirmation dialog.

Karaturhan, P., Orhan, İ., Kuşcu, K., & Yantaç, A. E. (2024). Informing the design of question-asking conversational agents for reflection. ***Personal and Ubiquitous Computing, 28*****(6), 1001–1019.** Karaturhan and colleagues study what happens when a conversational agent asks a question instead of giving an answer, and how the wording of that question changes whether a person actually stops to think. Their findings back the three-part structure of the Reflection Prompt in Section 10.3: observation, connection, and an open question, in that order, produced the most honest self-assessment from participants.

Mills, S., Whittle, R., Ahmed, R., Walsh, T., & Wessel, M. (2023). Dark patterns and sludge audits: an integrated approach. *Behavioural Public Policy, 1–27.* Mills and co-authors study the opposite of positive friction: sludge, the extra steps a company piles on to keep a user trapped in a bad decision, like a subscription that takes eight clicks to cancel. Their audit method sorts friction by who it protects. This gives Section 10.2 a clean test for the difference between a Guardian and a dark pattern: one slows the user down to protect them, the other slows the user down to trap them.

Ryan, R. M., & Deci, E. L. (2000). Self-determination theory and the facilitation of intrinsic motivation, social development, and well-being. *American Psychologist, 55*(1), 68-78. The authors establish autonomy as a fundamental human psychological requirement. People actively rebel when systems try to control them. This paper validates the absolute necessity of the master off switch. It demonstrates that users will only accept a proactive guardian if they maintain ultimate power to disable it.

Sheeran, P., & Webb, T. L. (2016). The intention-behavior gap. *Social and Personality Psychology Compass, 10*(9), 503-518. Sheeran and Webb map the exact psychological disconnect between what people plan to do and what they actually do. They outline why long-term goals frequently fail in the face of short-term temptations. This provides the scientific foundation for the Guardian Agent framework and the necessity of Reflection Prompts discussed in Section 10.3.

Shneiderman, B. (2022). *Human-Centered AI.* Oxford University Press. Shneiderman outlines a framework for building artificial intelligence that amplifies human abilities while ensuring strict human oversight. He argues forcefully for predictable systems with clear user controls. This text provides the ethical and architectural backing for the absolute necessity of the Master Off Switch detailed in Section 10.4.

Sterz, S., Baum, K., Biewer, S., Hermanns, H., Lauber-Rönsberg, A., Meinel, P., & Langer, M. (2024). On the quest for effectiveness in human oversight: ***Interdisciplinary perspectives. Proceedings of the 2024 ACM Conference on Fairness, Accountability, and Transparency (FAccT '24),*** **2495–2507.** Sterz and her co-authors, writing across law, philosophy, and computer science, argue that human oversight is not a switch you either have or lack. It is a capability that can be effective or merely decorative, depending on whether the person holding it has the time, the information, and the standing to actually use it. This raises the bar for Section 10.4: a Master Off Switch only counts if the user can find it, understand what it does, and reach it before the damage is done.

Thaler, R. H., & Sunstein, C. R. (2008). ***Nudge: Improving decisions about health, wealth, and happiness.*** **Yale University Press.** This is the foundational text for choice architecture. Thaler and Sunstein explain how passive environmental tweaks influence human behavior. Section 10.1 uses this text as a baseline to explain why active, agentic systems require an evolution from passive nudges to proactive guardianship.

Zhu, H., Wang, C., Xiao, Q., & Shen, H. (2026). Design principles for human-agent interaction. *arXiv:2606.20630* ***[cs.HC],*** **Carnegie Mellon University.** Zhu and colleagues argue that an agent should not be judged on task success alone, since it also interacts with, adapts to, and occasionally fails the person using it. Their 14 principles span the whole relationship: the first meeting, the middle of a task, the long run, and the moment something breaks. This one is a preprint, not yet peer-reviewed, so treat it as a directional pointer for Section 10.1 rather than settled research.

CHAPTER 11

Measuring Trust: Research Methods for Agentic Systems

In Chapter 10, we designed the Guardian Agent to protect the user's long-term goals. Now, we must prove those designs actually work. This chapter equips you with a specific user research toolkit adapted for agentic systems. You will learn to measure trust directly using two quantitative methods (the Delegation Test and Trust Scales) and two qualitative methods (the Wizard of Oz study and longitudinal Diary Studies).

For years, product teams relied on standard engagement metrics like daily active users or time on screen. Those numbers fail when evaluating autonomous systems. A user might open an application every day to double-check the agent's math, creating high engagement driven entirely by suspicion rather than confidence. To build genuine reliance, you need a completely different testing approach.

You need to know exactly why a tired nurse ignores an AI patient summary, or why a junior developer confidently accepts an automated database migration. We will isolate the specific variables that cause a human to hand over control. By the end of this chapter, you will be able to transform trust from a vague feeling into a manageable, measurable metric.

V. Yocco, *The Persuasive Agent*, Design Thinking,
https://doi.org/10.1007/979-8-8688-3317-5_11

11.1 You Can't Improve What You Don't Measure

There's an old management adage, often attributed to Peter Drucker (certainly not the first person to state this concept), that says, "You can't manage what you don't measure." In the world of product development, we live by this. We measure everything: conversion rates, daily active users, click-through rates, task completion times, server uptime. We live in a sea of dashboards and key performance indicators (KPIs).

We do this because measurement transforms a vague goal (We want a successful product) into a concrete, manageable process. It tells us if we are winning, losing, or treading water.

For most of the history of digital design, the things we needed to measure were relatively straightforward. But in the agentic era, our most important design goal is one of the softest, most subjective, and most famously difficult concepts to quantify: **trust**.

Measuring trust can feel like trying to measure the color of a mood or the weight of a feeling. Faced with this ambiguity, many teams fall back on what's easy to measure: engagement, retention, and number of interactions, then hope those numbers stand in for trust.

This is a dangerous and often misleading assumption. A user might engage with an agent every day because they are forced to by their workflow, all the while hating and distrusting it.

High engagement does not equal high trust. To build genuinely trustworthy agents, we must move beyond hope and proxy metrics. We must face the challenge head-on and find ways to measure trust directly.

This is the work of a user researcher. Measuring trust closes the loop of user-centered design: you form a hypothesis about what your users need, ship a design, then gather evidence from real people to confirm or correct it. And while it can be challenging, it is far from impossible. In my decades leading research teams, I've learned that even the most abstract human concepts leave behind tangible evidence. Trust is not a magical vapor. It is a psychological state that manifests in a user's attitudes and, more importantly, in their behaviors. Our job is to become detectives, to find that evidence and use it to guide our design decisions.

This chapter is a practical guide to the craft of trust measurement. It is a UX researcher's toolkit, adapted specifically for the unique challenges of agentic AI. We will not be covering a dozen different, complex academic methods. Instead, we will focus on four robust, flexible, and immediately applicable techniques, two quantitative and two qualitative, that you can start using with your team tomorrow. These methods will allow you to move from guessing to knowing, and from hoping to measuring.

Table 11-1 maps the six methods this chapter draws on. It contrasts the two baseline approaches most teams already run, standard analytics and usability testing, with the four trust-focused methods we will build. Now, let us begin by looking at how we can quantify these behaviors and attitudes using our two quantitative methods.

Table 11-1. *Research methods for agentic systems*

Method Name	Data Category	What It Measures	Ideal Use Case	The Agentic Application
Standard Analytics (Baseline)	Quantitative	Raw engagement and clicks.	Tracking daily active users.	Identifying frequent drop-offs during an onboarding flow.
Usability Testing (Baseline)	Qualitative	Interface friction.	Testing button placement.	Observing a user click the wrong settings toggle.
The Delegation Test	Quantitative	Behavioral reliance.	Calculating the exact point a user takes back control.	Counting how many users let the agent send an external email.
Trust Scales	Quantitative	User attitudes and perception.	Scoring competence and integrity.	Surveying users immediately after an AI interaction.
Wizard of Oz	Qualitative	Conversational dynamics.	Prototyping personas early.	Having a human simulate the agent's replies in real time.
Diary Studies	Qualitative	Long-term relationship health.	Tracking attitude changes over weeks.	Asking users to record audio memos in their natural environment.

11.2 Quantitative Methods: The Delegation Test and Trust Scales

Quantitative methods are about answering the question, what is happening? They produce numbers, charts, and statistically significant results. They are excellent for tracking changes over time, comparing different design variations (A/B testing), and understanding the scope of a problem.

The Delegation Test: Measuring Trust Through Behavior

The single most reliable way to measure trust is to observe behavior. A user can *say* they trust your agent, but the real proof is whether they are willing to *act* on that trust. Are they willing to delegate a meaningful task? The Delegation Test is a simple, powerful behavioral experiment designed to measure exactly that.

The concept builds directly on the Delegation Ladder we designed in Chapter 7. Instead of waiting for a user to climb that ladder organically, we present them with a series of tasks of increasing risk and see how high they are willing to climb in a controlled setting.

How to Run a Delegation Test

1. **Define Your Ladder:** Create a series of 5–7 tasks that a user could perform with your agent. These tasks should represent a clear escalation of risk or consequence. Risk can mean financial cost, time commitment, social consequence, or irreversibility.

2. **Present the Choice:** For each task on the ladder, you will present the user with a clear choice:

    - **Option A:** Perform the task yourself using the traditional, manual interface.

    - **Option B:** Delegate the task to the AI agent.

3. **Measure the Crossover Point:** The key metric you are looking for is the Delegation Crossover Point, as shown in Figure 11-1; this is the first rung on the ladder where the user chooses to perform the task manually rather than delegating it to the agent. This is the precise point where their trust ran out.

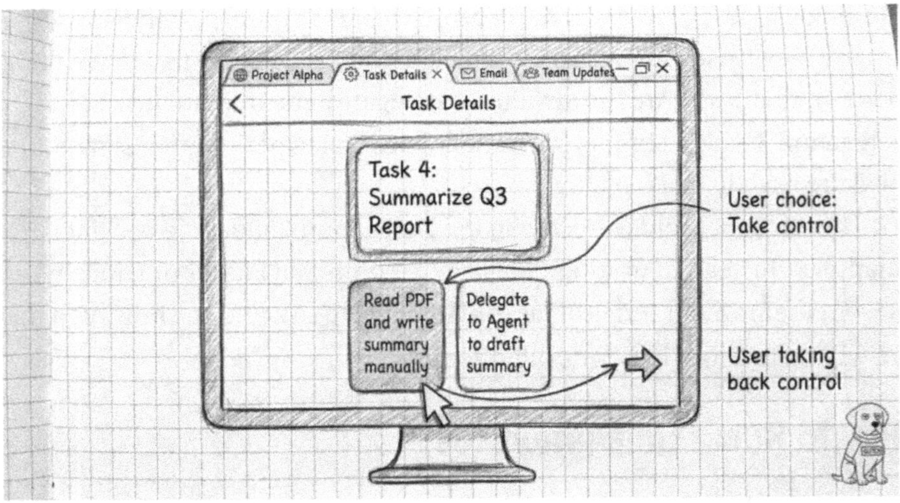

***Figure 11-1.** The Delegation Crossover Point. A user chooses to perform a high-stakes task manually, indicating the exact moment their behavioral trust runs out. Image generated by author using Gemini Pro 3.1/Nano Banana 2*

Let's imagine we're testing our Travel Companion agent, the flight-booking assistant we first met in Chapter 1. Our Delegation Ladder test might look like this:

Task 1 (Lowest Risk): You have 15 hotels saved from a search for your upcoming trip. Would you like to

A. Manually sort them into a shortlist yourself

B. Let the agent group them by neighborhood and price for you

Task 2: A colleague has invited you to co-present at a conference next month. Would you like to

A. Check your calendar and reply to the invitation yourself

B. Let the agent check your calendar and confirm your attendance for you

Task 3: Your connecting flight has been delayed, and you will now arrive at your hotel four hours later than planned. Would you like to

A. Call the hotel yourself to let them know

B. Let the agent message the hotel with your new arrival time

Task 4: You need to book a flight to your quarterly meeting in Denver, and the agent has narrowed it down to a direct outbound flight and a return flight with a one-hour layover that saves $120. Would you like to

A. Compare the two flights yourself and book the one you prefer

B. Let the agent book the itinerary it recommends, using your corporate card

Task 5 (Highest Risk): Your connecting flight has just been canceled due to weather, and you must reach Denver in time for the morning session. Would you like to

A. Call the airline yourself and manage the rebooking

B. Let the agent automatically rebook you on any available flight, even if it costs significantly more, to guarantee you make the meeting

By running a group of users through this test, you get a hard behavioral metric. You might find that with Design A the median crossover point sits at rung 2, with most users refusing to delegate the delayed-arrival message, while a firmer, higher-authority persona in Design B pushes the median to rung 4. Comparing the two distributions side by side shows you exactly where the new persona earned deeper delegation, and where users still drew the line.

Scoring the Delegation Crossover Point

Number the tasks on your ladder from 1 (lowest risk) to N (highest risk). For a single participant, record the rung of the first task where they pick the manual option (Option A) instead of delegating (Option B). That rung is their Delegation Crossover Point (DCP). A participant who delegates every rung never crosses over, so score them N + 1. The result sits on a fixed range from 1 to N + 1, where a higher number means the participant delegated deeper into the ladder before pulling back. A companion figure, delegation depth, is DCP minus 1: the count of consecutive rungs a person handed over before their first refusal.

The single number that matters most for each person is the last rung they were willing to delegate, so start there. Plot the full distribution of crossover points across your participants and report the median, not the mean alone. The mean can mislead. If half your users delegate everything and half refuse at rung 2, the average lands near the middle, at a rung almost nobody actually chose. Use the mean only when you compare two designs whose distributions have a similar shape, such as an A/B test where both curves are roughly bell-shaped. When the distribution is bimodal or skewed, the median and the histogram tell the honest story.

Worked example (five-rung Travel Companion ladder, one participant who delegates rungs 1 through 3, then goes manual at rung 4):

- Rung 1, sort saved hotels: delegated. Still delegating.
- Rung 2, confirm conference RSVP: delegated. Still delegating.
- Rung 3, notify hotel of delay: delegated. Still delegating.
- Rung 4, book the Denver flight: went manual. First refusal, so Delegation Crossover Point = 4.
- Rung 5, rebook a canceled connection: went manual. Delegation depth = 3.

Trust Scales: Measuring Trust Through Attitude

While behavior is the gold standard, it doesn't tell the whole story. A user's internal attitudes, beliefs, and feelings about the agent are also critically important. The best way to measure these is with a **Trust Scale**, a type of survey that uses a series of statements to quantify a user's perceptions.

The most common format is the Likert scale, where a user rates their agreement with a statement, typically on a 5- or 7-point scale (e.g., from "Strongly Disagree" to "Strongly Agree").

Academics have developed many complex, validated trust scales. For our practical purposes, we can create a simple, effective scale by focusing on the three core pillars of perceived trustworthiness:

1. **Competence/Ability:** Does the user believe the agent has the skills and knowledge to do its job?
2. **Benevolence:** Does the user believe the agent is on their side and has their best interests at heart?
3. **Integrity:** Does the user believe the agent is honest and reliable in its principles?

A Sample Nine-Item Trust Scale

Please rate your agreement with the following statements on a scale of 1 (Strongly Disagree) to 5 (Strongly Agree).

- **(Competence)** The agent is effective at its job.
- **(Competence)** The agent seems knowledgeable in its domain.
- **(Competence)** The agent makes very few errors.
- **(Benevolence)** The agent acts in my best interest.
- **(Benevolence)** I feel like the agent is on my side.
- **(Benevolence)** The agent's suggestions are genuinely helpful.
- **(Integrity)** The agent is reliable and consistent.
- **(Integrity)** The agent is honest about its limitations.
- **(Integrity)** I can depend on the agent to behave as expected.

By having users complete this survey after an interaction, you can calculate an overall "Trust Score" (e.g., the average of all 9 ratings). This score is incredibly useful for A/B testing. Does a more humorous persona increase the Benevolence score but decrease the Competence score? Does adding an "explainability on demand" feature increase the Integrity score? This simple survey gives you the diagnostic data you need to understand the attitudinal impact of your design choices.

You can also split this score by segment to test persona fit directly. Run persona A with one group and persona B with another, then compare their Benevolence and Competence averages side by side. Add a warmth item as well, such as the agent feels approachable and considerate, because a persona can rate high on competence and still feel cold to the audience you care about.

With our quantitative baseline established, we now turn to qualitative methods to uncover the human stories behind those numbers.

11.3 Qualitative Methods: The "Wizard of Oz" Study for Agents

Quantitative methods tell you *what* is happening. Qualitative methods tell you *why*. They are about the rich, messy, human stories behind the numbers. For agentic systems, the single most powerful qualitative method is one that allows you to test your agent's persona and conversational abilities long before you write a single line of AI code. It's called the **Wizard of Oz (WoZ) study**.

The concept is brilliantly simple. You create a prototype of the agent's interface. A real user interacts with this interface, thinking they are talking to a sophisticated AI. But behind the scenes—behind the curtain, like the Wizard of Oz—a human researcher (the "wizard") is intercepting the user's inputs and typing back the agent's responses, simulating the behavior of the AI.

Why is this so powerful?

- **It's Fast and Cheap:** You can test a new agent persona or a complex conversational flow in a single afternoon, without any engineering investment.
- **It Separates the Persona from the Tech:** WoZ allows you to test the *ideal* user experience. If users are still frustrated when interacting with a "perfect," all-knowing human wizard, then you know the problem isn't with your AI model; it's with your core interaction design.

- **It's a Prototyping Tool for Personality:** This is the ultimate way to workshop your agent's character. Is the "Thrifty Grandparent" persona coming across as wise and helpful, or just old-fashioned and condescending? You can have the wizard try different tones of voice, different levels of proactivity, and see how users react in real time.

Run the same scenario with two distinct user groups and watch where their reactions split. A time-pressed emergency-room nurse and a curious retiree can meet the same Thrifty Grandparent persona and walk away with opposite verdicts. When one segment relaxes and the other bristles, you have found evidence that persona fit depends on audience, exactly as we argued in Chapter 5.

How to Run a Wizard of Oz Study

1. **Create the Puppet:** Build a simple, high-fidelity prototype of the agent's UI. This is what the user will see. It needs to look real.
2. **Set Up the Control Room:** The wizard needs an interface where they can see the user's input in real time and type back a response that will appear in the user's UI. This can be done with specialized software, or even a cleverly hacked Google Doc. Figure 11-2 shows an example of how a control room setup might function.
3. **Define the Agent's Rules:** The wizard cannot just act like a human. They must follow the rules defined in the **Agent Persona Canvas**. What is the agent's

vocabulary? What are its known limitations? The wizard's job is to *role-play* the agent as faithfully as possible.

4. **Run the Session:** Bring in a real user. Give them a task to complete with the "agent." As they type or speak their prompts, the wizard, in another room, types back the agent's response according to the rules. An observer should be with the user, prompting them to think aloud ("What are you expecting the agent to do now? How did that response make you feel?").

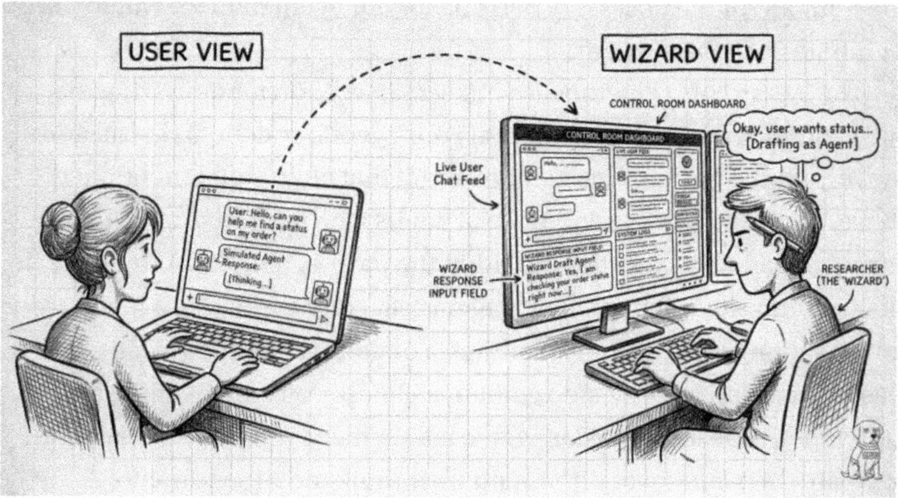

Figure 11-2. *The Wizard of Oz Setup. A human researcher simulates the agent's responses, allowing teams to test conversational design without writing any AI code. Image generated by author using Gemini Pro 3.1/Nano Banana 2*

I once ran a WoZ study for a Travel Companion agent, similar to the flight-booking assistant from Chapter 1. We thought a hyper-efficient, "Just the facts" persona would be best. Within ten minutes, our first participant was visibly frustrated. "It feels like I'm talking to a robot," she said. "It's giving me the flight times, but it's not helping me *decide*."

For the next participant, we told the wizard to switch personas on the fly to a more collaborative Travel Companion. The wizard started using "we" language ("Okay, let's see what we can find") and acknowledging the user's stress ("Planning a big trip can be overwhelming, let's take it one step at a time").

The change was instantaneous. The user relaxed, started collaborating, and praised how smart and helpful the agent was. We learned more in that one hour about the necessary persona for our agent than we would have in six months of development.

The Wizard of Oz method is your secret weapon for de-risking the most complex and uncertain part of your agentic system: its social and conversational design. One caution about that afternoon I mentioned above, we tested a single audience of adults who had children in their household and had vacationed within the past six months. A frequent business traveler might prefer an agent reflecting speed and a terse confirmation, while a family planning a once-a-year holiday wants reassurance and step-by-step guidance. Before you crown one persona the winner, repeat the session with each segment you plan to serve and compare their crossover points and their think-aloud sentiment.

Your segment comparison will point to one of two design paths. If the groups want the same core character with small tonal changes such as formality or verbosity, build one adaptive persona that reads context and adjusts. If the groups want fundamentally different characters, a cautious Expert Analyst for one and a warm Companion for another, ship several fixed personas and let users pick during onboarding. Let the study data settle this question, not a whiteboard debate.

While a Wizard of Oz study provides immediate feedback on initial interactions, we must also track how users behave over weeks or months. Let us examine diary studies to capture that long-term reality.

11.4 The Diary Study: Tracking Trust Over Time

Lab studies offer a controlled environment. You sit across from a participant, watch them click through a prototype, and record their immediate reactions. This works well for basic usability. It fails completely when evaluating an ongoing relationship. Trust builds and breaks over days, weeks, and months. A 60-minute session captures a single photograph. We need a full motion picture.

A diary study solves this problem. This qualitative method requires participants to log their experiences in their natural environment as they happen. You ask a user to record a short entry every time they interact with the AI agent. They might send a text message from a noisy cafe, fill out a quick form on a commuter train, or record a short voice memo at their office desk. This method captures reality.

Agentic systems act on behalf of the user. To hand over control, a human needs consistent proof of reliability. A user might confidently delegate a complex spreadsheet formatting task on Monday. On Tuesday, the agent makes a minor error. By Wednesday, the user stops using the feature entirely. If you test this user on Monday, you record a total success. A diary study catches the failure on Tuesday and the abandonment on Wednesday. It exposes the exact moment a relationship fractures.

You must structure these studies carefully to avoid fatiguing your participants. Give them a simple trigger. Ask them to pull out their phone and answer two specific questions immediately after the agent completes a specific action. Keep the friction low. You want raw, immediate emotion. Did the agent save them time? Did a specific UI pattern cause confusion? Did they feel comfortable letting the machine draft an invoice for a major client?

At the end of a two-week period, you collect hundreds of these small entries. You look for patterns. You trace the exact trajectory of trust. You see exactly when a user decides to flip the master off switch we designed in Chapter 10. You gather a clear, documented timeline of human behavior. This qualitative data gives your design team the exact evidence they need to fix broken interactions.

Example Diary Study: Understanding Trust in a New AI-Powered Financial Advisor

Product: FinBot, an AI-powered financial advisor that offers personalized investment recommendations and budget planning.

Goal of Study: To understand how users develop (or lose) trust in FinBot over a 3-week period in their natural financial management routines.

Participants: Ten individuals (five early adopters of financial tech, five more traditional investors), all with active investment portfolios and/or budgeting needs.

Diary Setup: A private Slack channel where participants are invited to post daily entries. Each participant has a unique thread to maintain privacy and facilitate focused follow-up.

Duration: 3 weeks-----**Week 1: Initial Impressions and Onboarding Trust**

Daily Prompts:

- "Describe your main financial goal today (e.g., check portfolio, pay bills, review budget). Did you use FinBot to help you achieve it? If so, how? What was the outcome?"
- "What was your initial reaction to FinBot's suggestions today? Did you follow them? Why or why not?"

- "Was there anything FinBot did today that felt particularly helpful or particularly frustrating? Explain."
- "On a scale of 1–5, how transparent do you feel FinBot is about *why* it's making a certain recommendation?"
- "Screenshot and share any interaction that made you question FinBot's intelligence or reliability."

Anticipated Insights:

- Initial user skepticism vs. immediate acceptance
- Ease of onboarding and understanding FinBot's capabilities
- Early "trust injuries" if FinBot makes a mistake or gives confusing advice
- Identification of common first-week use cases and points of friction

-----Week 2: Developing Habits and Handling Complexity
Daily Prompts:

- "Please describe one significant financial decision you faced today (e.g., considering a new investment, making a large purchase). Did FinBot play a role? How did its advice compare to your own thoughts or other sources?"
- "Did you encounter a situation today where you *could have* used FinBot for advice, but chose not to? What was your reasoning?"
- "Tell us about a time today FinBot's advice surprised you (positively or negatively). How did it impact your trust?"

- "On a scale of 1–5, how much do you feel FinBot *understands* your personal financial situation and goals?"
- "If FinBot made a recommendation you disagreed with, how did you handle it? Did you trust your own judgment more, or did you try to understand FinBot's reasoning?"

Anticipated Insights:

- Whether users are integrating FinBot into their regular financial routines.
- How users validate or challenge FinBot's advice on more complex matters.
- The impact of FinBot's explanations on user trust.
- Emergence of "trust barriers" in specific financial scenarios.

-----Week 3: Long-Term Reliance and Stress Testing Trust

Daily Prompts:

- "Reflect on a moment this week where you felt truly reliant on FinBot. What happened, and how did it feel?"
- "Was there a high-stakes financial situation you faced today? How did FinBot perform, and how did its performance affect your overall trust?"
- "If you were to recommend FinBot to a friend, what would be the single most important reason? What, if anything, would be a caution?"

- "On a scale of 1–5, how confident are you in FinBot's ability to help you achieve your long-term financial goals?"
- "Looking back over the past three weeks, describe one specific interaction that significantly *built* your trust in FinBot, and one that significantly *eroded* it (if any)."

Anticipated Insights:

- The point at which users feel comfortable delegating significant financial tasks to FinBot
- How FinBot handles critical errors or perceived failures, and the user's recovery from "trust injuries"
- Identification of power users vs. those who remain skeptical
- Qualitative stories that illustrate the full arc of the user-agent trust relationship

-----**Data Analysis:**

- **Qualitative Coding:** Transcribe and code all diary entries for themes related to trust, skepticism, satisfaction, frustration, understanding, and perceived competence.
- **Trust Trajectories:** For each participant, plot a "trust journey" based on their daily trust scores and qualitative feedback, identifying key moments of increase or decrease in trust.
- **Critical Incidents:** Identify specific interactions (positive and negative) that significantly impacted user trust.

- **Segment Comparison:** Split your trust trajectories by the groups you recruited, the early adopters and the traditional investors, and look for divergence. If the two lines climb and fall together, one persona serves both. If they pull apart, you have quantitative grounds for tailoring a different persona to each segment.
- **Language Analysis:** Pay attention to the language users employ when describing their interactions with FinBot (e.g., "it understood," "it failed me," "I had to double-check").

This detailed diary study would provide a rich, longitudinal understanding of how trust in an AI-powered financial advisor develops, solidifies, or breaks down in real-world usage, offering invaluable insights for product improvement and trust-building strategies.

These four methods provide a complete toolkit for evaluating user reliance. We can now summarize how to integrate them into your workflow.

11.5 Summary

- Measuring trust is not optional. To manage and improve the trustworthiness of our agents, we must move beyond proxy metrics and find ways to measure trust directly.
- A robust measurement strategy uses a mix of quantitative (what is happening) and qualitative (why it's happening) methods.
- **Quantitative methods** give us the numbers. The **Delegation Test** provides a behavioral metric of how much risk a user is willing to delegate. **Trust Scales** provide an attitudinal metric of the user's perceptions of the agent's competence, benevolence, and integrity.

- **Qualitative methods** give us the stories. The **Wizard of Oz study** is a powerful, low-cost method for prototyping and testing an agent's persona and conversational design before it's built. The **Diary Study** is an essential longitudinal method for understanding how the user-agent relationship and trust evolve over time in a real-world context.
- Trust is not an unknowable, magical quality. It is a designable, measurable, and manageable outcome. By equipping ourselves with these research methods, we can move from designing in the dark to iterating our way toward agents that are genuinely worthy of their users' trust.

Table 11-2 pulls the four core methods together so you can pick the right one for the quest-ion in front of you.

- **Delegation Test**: Question it answers: will users actually hand over the task? Data type: quantitative, behavior. Reach for it when: you need proof of reliance rather than opinion, or want to compare two designs. Output: crossover-point distribution.
- **Trust Scales**: Question it answers: what do users believe about the agent? Data type: quantitative, attitude. Reach for it when: you want to track competence, benevolence, and integrity after an interaction. Output: trust score, 1 to 5.
- **Wizard of Oz**: Question it answers: does the persona and conversation work? Data type: qualitative. Reach for it when: you want to test the ideal experience before building the model. Output: observed reactions and quotes.

- **Diary Study**: Question it answers: how does trust move over time? Data type: qualitative, longitudinal. Reach for it when: you need the trajectory of a real relationship across weeks. Output: a trust timeline per user.

Table 11-2. *Choosing a trust-measurement method by the question you need answered*

Research Method	Question It Answers	Data Type	When to Use	Output
Delegation Test	Will users actually hand over the task?	Quantitative, behavior	When you need proof of reliance rather than opinion, or want to compare two designs.	Crossover-point distribution.
Trust Scales	What do users believe about the agent?	Quantitative, attitude	When you want to track competence, benevolence, and integrity after an interaction.	Trust score, 1 to 5.
Wizard of Oz	Does the persona and conversation work?	Qualitative	When you want to test the ideal experience before building the model.	Observed reactions and quotes.
Diary Study	How does trust move over time?	Qualitative, longitudinal	When you need the trajectory of a real relationship across weeks.	A trust timeline per user.

In Chapter 12, we will apply this entire analytical framework to deconstruct four realistic case studies, observing how these design principles hold up when a user attempts to book a flight to Italy or summarize a team meeting.

11.6 Chapter 11 Workshop: The Trust Measurement Plan

In this chapter, we moved from the design studio into the research lab. We established that trust, while complex, is not a mysterious vapor. It is a measurable human response. We've added four powerful research methods to our toolkit for capturing both the "what" and the "why" of our users' trust.

This workshop is your first field test. You will take on the role of a user researcher and create a comprehensive "Trust Measurement Plan" for the agent you have been designing. This is the process that separates wishful thinking from evidence-based design. It is how we prove that our well-intentioned designs are having the impact we desire.

Part 1: Your Quantitative Toolkit

First, we will design the tools to get the numbers. This is how we will measure and track trust at scale. Pull out your **Agent Persona Canvas** and the other design documents for your agent project.

1. **Design a 3-Rung Delegation Test:** The Delegation Test is the ultimate measure of behavioral trust. Your task is to design a simplified, 3-rung version of this test, defining a clear escalation of risk for your specific agent.

   - **Rung 1—The Low-Risk Delegation**: What is a simple, easily reversible task the user could delegate?
     - *Example (for a Thrifty Grandparent recipe agent):* "The agent has found three ways to use up your leftover roast chicken. Do you want to A) Read all three recipes yourself and pick one, or B) Let the agent pick the one it thinks best matches your taste and start walking you through it?"
   - **Rung 2—The Medium-Risk Delegation**: What is a task that has a higher cost if the agent gets it wrong (e.g., social risk, moderate time/money)?
     - *Example:* "The recipe needs three ingredients you don't have. Do you want to A) Write them down and add them to your own shopping list, or B) Let the agent add them directly to your grocery delivery app's cart?"
   - **Rung 3—The High-Risk Delegation**: What is a high-stakes, potentially irreversible task that would require a significant leap of faith to delegate?
     - *Example:* "It's your in-laws' anniversary dinner tonight, and the agent thinks a change of menu would vimpress them more. Do you want to A) Review its suggested menu change

yourself before deciding, or B) Let the agent automatically update the shared family recipe box so everyone sees the new plan?"

2. **Design Your Custom Trust Scale**: Now, let's design the tool to measure attitudes. Using the three pillars of trustworthiness, write three custom survey questions for a 5-point Likert scale (Strongly Disagree to Strongly Agree). These should not be generic; they must be tailored to your agent's specific persona and function.

- **One Competence Question:**
    - *Generic:* "The agent is effective."
    - *Your Custom Question: [Write a specific question. Example for a "Thrifty Grandparent" recipe agent: "I am confident that the agent can create a tasty meal out of the ingredients I give it."]*
- **One Benevolence Question:**
    - *Generic:* "The agent is on my side."
    - *Your Custom Question: [Write a specific question. Example: "I feel like the agent genuinely wants to help me reduce my food waste and save money."]*
- **One Integrity Question:**
    - *Generic:* "The agent is reliable."
    - *Your Custom Question: [Write a specific question. Example: "When the agent says it doesn't know a fancy recipe, I believe it and trust its honesty."]*

You now have a behavioral test and an attitudinal survey. This is the core of your quantitative measurement plan.

Part 2: Your Qualitative Toolkit

The numbers tell you what's happening, but the stories tell you why. Now, we'll design the studies to uncover those crucial narratives.

1. **Plan Your Wizard of Oz "Critical Test"**: The WoZ study is your persona-testing laboratory. Imagine you are about to run a WoZ study for your agent. You don't need to design the whole protocol, just the most important part.
    - **Identify a Critical Scenario:** What is the one make-or-break interaction that will test the core of your agent's persona? What is the user likely to do or say that will put your design choices to the test?
        - *Example (for the Thrifty Grandparent recipe agent):* "The user confesses to the agent that they let an entire bag of spinach go bad again this week and feels guilty, like they are bad at this whole reducing-food-waste thing."
    - **Write the Wizard's Script:** Now, write the *exact* response the wizard should give in that scenario, following the rules of your Agent Persona Canvas. This is the "ideal" response you want to test.
        - *Example Script:* "Oh, don't you fret about that, dear. Spinach is a slippery one, it goes fast whether you're watching it or not. Let's not dwell on it. Why don't we take a look at what's

left in your fridge right now and find something good to do with it before it follows the spinach out the door?"

This exercise forces you to translate your abstract persona into a concrete conversational strategy for a high-stakes, emotional moment.

2. **Design Your Trust Tracker Diary Prompts**: Finally, let's plan for the long term. A diary study is how you track the evolution of the user-agent relationship. The key is asking the right questions.

   Your task is to write **two powerful, open-ended diary prompts** to send to your participants each day.

   - **Prompt 1: The Trust Building Prompt.** Write a question designed to capture a moment where trust *increased.*

     - *Example:* Was there a moment today where the agent did something unexpectedly smart, helpful, or considerate? Describe what happened and how it made you feel about the agent.

   - **Prompt 2: The Trust Barrier Prompt**: Write a question designed to uncover a moment where trust was *broken* or *never attempted.*

     - *Example:* Think back on your day. Was there a task or a moment where you *could* have used the agent, but you chose to do it yourself instead? Walk me through that moment. What was the task, and why did you decide not to delegate it?

This second question is a secret weapon. It is often more insightful to find out why a user *didn't* use your agent than why they did. It reveals the hidden fears and trust barriers that a simple usability test will never uncover.

You have now created a complete, multi-method Trust Measurement Plan. You have a behavioral test to measure what users *do*, a survey to measure what they *think*, a qualitative script to test your persona's ideal state, and longitudinal prompts to track how the relationship evolves over time. This is no longer a soft concept. You have made trust measurable.

11.7 References and Further Reading

Bo, J. Y., Wan, S., & Anderson, A. (2025). To rely or not to rely? Evaluating interventions for appropriate reliance on large language models. *Proceedings of the 2025 CHI Conference on Human Factors in Computing Systems,* Article 905. This study tests design interventions that move users toward appropriate reliance rather than blanket acceptance of an LLM's output. It supports the Delegation Test as a way to read whether a design earned deeper delegation or pushed users into over-reliance, directly relevant to Section 11.2.

Bolger, N., Davis, A., & Rafaeli, E. (2003). Diary methods: Capturing life as it is lived. *Annual Review of Psychology, 54*(1), 579-616. This comprehensive review outlines the exact mechanics of diary studies. The authors explain how recording data in a natural environment eliminates the memory errors common in retrospective surveys. This paper provides researchers with a rigid framework for structuring the daily prompts discussed in Section 11.4.

Carter, S., & Mankoff, J. (2005). When participants do the capturing: the role of media in diary studies. *Proceedings of the SIGCHI Conference on Human Factors in Computing Systems,* 899-908. Carter and Mankoff examine how users record information during longitudinal studies. They evaluate different input methods, from text to audio. This research helps teams lower the friction of participation, ensuring users actually complete their diary entries over a multi-week test.

Dzindolet, M. T., Peterson, S. A., Pomranky, R. A., Pierce, L. G., & Beck, H. P. (2003). The role of trust in automation reliance. *International Journal of Human-Computer Studies*, 58(6), 697-718. These researchers map the exact relationship between perceived trustworthiness and actual usage. They demonstrate that users often abandon automated systems after observing a single error. This study directly supports the necessity of the Delegation Test to measure behavioral drop-off points.

Fang, J., Arechiga, N., Namikoshi, K., Bravo, N., Hogan, C., & Shamma, D. A. (2024). On LLM wizards: Identifying large language models' behaviors for Wizard of Oz experiments. *Proceedings of the 24th ACM International Conference on Intelligent Virtual Agents (IVA '24)*. This paper offers a modern update to Wizard of Oz practice, examining when a language model can stand in for the human wizard and where that substitution fails. It gives Section 11.3 a current empirical anchor for running WoZ studies in the era of capable language models.

Guo, Z., Wu, Y., Hartline, J. D., & Hullman, J. (2024). A decision theoretic framework for measuring AI reliance. *Proceedings of the 2024 ACM Conference on Fairness, Accountability, and Transparency (FAccT '24)*, 221-236. This paper offers a formal way to separate reliance from a rational baseline, which grounds the crossover-point scoring method added to Section 11.2 in current measurement theory.

Gutzwiller, R. S., Yousefi, R., Larson-Calcano, T., Lee, J. R., Verma, A., Tenhundfeld, N. L., & Maknojia, I. (2025). Systematic review of the use and modification of the Trust in Automated Systems Scale. *Proceedings of the Human Factors and Ergonomics Society Annual Meeting*. This review examines 1,480 uses of the Jian et al. trust scale and shows that altering wording or item counts often lowers validity. It is a practical warning for teams building the custom Trust Scale described in Section 11.2.

He, G., Demartini, G., & Gadiraju, U. (2025). Plan-then-execute: An empirical study of user trust and team performance when using LLM agents as a daily assistant. *Proceedings of the 2025 CHI Conference on Human Factors in Computing Systems*, 1-22. This recent field-style study examines how people trust and rely on an LLM agent handling everyday tasks. It gives Section 11.1 a current empirical anchor for the claim that engagement and trust diverge.

Hoff, K. A., & Bashir, M. (2015). Trust in automation: Integrating empirical evidence on factors that influence trust. *Human Factors*, *57*(3), 407-434. Hoff and Bashir synthesize decades of data into a unified model of human-machine reliance. They categorize the specific variables that cause a person to trust an algorithm. This text acts as a complete reference guide for teams designing the qualitative questions in their attitudinal surveys.

Jian, J. Y., Bisantz, A. M., & Drury, C. G. (2000). Foundations for an empirically determined scale of trust in automated systems. *International Journal of Cognitive Ergonomics*, *4*(1), 53-71. This paper documents the creation of a standardized, validated survey for evaluating machines. The authors systematically identify the exact words people use to describe reliable systems. You can use their validated questions as a direct template for the Trust Scales introduced in Section 11.2.

Mayer, R. C., Davis, J. H., & Schoorman, F. D. (1995). An integrative model of organizational trust. *Academy of Management Review*, *20*(3), 709-734. Mayer, Davis, and Schoorman established the three pillars of trustworthiness: ability, benevolence, and integrity. While originally applied to human organizations, this framework directly translates to artificial agents. This paper provides the foundational theory for the custom survey questions in the chapter workshop.

Patton, C. E., & Wickens, C. D. (2024). The relationship of trust and dependence. *Ergonomics, 67*(11), 1535-1552. This paper draws a clean line between trust, an attitude, and dependence, a behavior, which is exactly the split between the Trust Scale and the Delegation Test. It is a tidy framing citation for the opening of Section 11.2.

Riek, L. D. (2012). Wizard of Oz studies in HRI: a systematic review and new reporting guidelines. *Journal of Human-Robot Interaction, 1*(1), 119-136. Riek reviews years of simulated interaction studies and identifies common methodological flaws. She provides a strict checklist for running a simulation accurately. This modern update ensures your team builds a rigorous testing environment when evaluating a new agent persona.

Steyvers, M., & Kumar, A. (2024). Three challenges for AI-assisted decision-making. *Perspectives on Psychological Science, 19*(5), 722-734. This paper documents the gap between what a model knows and what users believe it knows, reinforcing why proxy metrics fail and why direct measurement of trust matters, as argued in Section 11.1.

CHAPTER 12

Persuasion in the Wild: Case Studies and Future Directions

Our journey is nearly at its end. We have spent 11 chapters assembling a powerful and sophisticated toolkit for understanding, designing, and measuring trust in agentic systems. We have our frameworks, our principles, and a compass for making responsible choices. Now, it is time to use them.

In this final chapter, we will do what all good researchers and designers do: we will observe the world as it is. We will step into the role of a critic and apply our entire analytical framework to a series of hypothetical but realistic case studies. We will deconstruct these agents, celebrate what they do well, and diagnose their failures with the precision our new vocabulary allows. This is the final exam. Let's see how our tools hold up in the wild.

V. Yocco, *The Persuasive Agent*, Design Thinking,
https://doi.org/10.1007/979-8-8688-3317-5_12

12.1 Deconstructing a Shopping Agent

Case Study 1: ThreadFinder, the AI Personal Stylist

Product: ThreadFinder is a feature within a large online fashion retailer. Its stated goal is to act as a personal stylist, helping users discover new items and build outfits from the retailer's vast catalog.

First Interaction: A new user, Maya, opens the feature for the first time. The screen shows a friendly avatar and a message:

> "Welcome to ThreadFinder! I'm here to be your personal AI stylist. To get started, let's figure out your vibe. Here are five different styles. Tap on the ones you like!"

Analysis: We begin our deconstruction at the very beginning, with the first impression.

- **Persona:** The agent immediately establishes its persona. The archetype is a **Friendly Personal Stylist.** The language ("vibe," "let's") is casual and collaborative, aiming to build **Liking** through a feeling of partnership. So far, so good.
- **Reciprocity:** Here we see our first missed opportunity. Figure 12-1 shows the exact moment the agent forces a user into a rigid style box instead of observing natural browsing behavior. The agent's first move is to ask the user to do work ("Tap on the ones you like!"). It has not yet offered a proactive gift of competence. A better approach might have been to analyze Maya's past purchases and say, "Welcome! I've taken a look at your past orders. It seems you're a fan of minimalist

styles and neutral colors. Is that right?" This would have been a powerful gift of being understood, immediately triggering reciprocity.

- **Consistency and the Delegation Ladder:** The initial request to tap on styles is a perfect **Rung 1** on a Delegation Ladder. It's a small, low-risk commitment. Maya is simply providing data, not delegating a task. This is a classic foot-in-the-door technique.

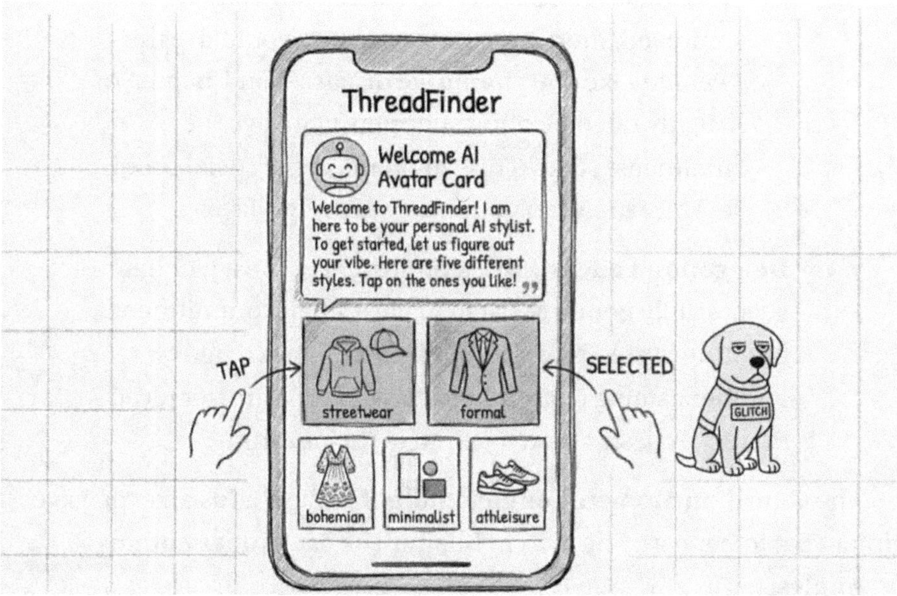

Figure 12-1. *The Premature Categorization. Image generated by author using Gemini Pro 3.1/Nano Banana 2*

Core Interaction: Maya taps on a few styles. The agent then presents her with a curated page of individual clothing items.

> "Great choices! Based on your selections, here are some items I think you'll love."

Below each item is a star rating and the text "Trending this week."

Analysis:

- **Social Proof:** The agent is using two forms of social proof here. The star rating is a classic, though blunt, tool. The "Trending this week" badge is a form of Scarcity (specifically, a time-based urgency) combined with social proof. However, it's generic. It doesn't leverage the agent's unique intelligence.
    - **Diagnosis:** The social proof is not contextual. An improved, more persuasive version would use the "People Like You" frame we discussed in Chapter 7: "Highly rated by other shoppers who love the minimalist style." This would make the social proof feel more relevant and trustworthy to Maya.
- **Delegation Ladder (Stuck on Rung 1):** The agent has successfully gotten Maya to make a small commitment, but its next step is weak. It has simply shown her another catalog page. It hasn't asked for a greater level of trust. It hasn't tried to climb to the next rung.

The Guardian Moment (or lack thereof): Maya adds an expensive winter coat to her cart. She lives in Miami. The agent does nothing.

Analysis:

- **Guardian Agent Failure:** This is a critical failure of guardianship. The agent has powerful contextual information (Maya's location) that suggests this purchase might be a mistake or an impulse buy. This is a perfect opportunity for a **Reflection Prompt**.

- **The Missed Opportunity:** A Guardian Agent would have intervened with a gentle, non-judgmental prompt: "(Observation) That's a beautiful coat! (Connection) I notice you're shipping to a Miami zip code, where it's currently 85 degrees. (Question) Are you planning a trip somewhere colder, or should we save this to a 'Winter Wishlist' for later?" This intervention is not about blocking the sale. It's about preventing a likely return and a moment of post-purchase regret. It is an act that would build profound, long-term trust, even at the cost of a potential short-term conversion.

Final Verdict on ThreadFinder:

ThreadFinder has a decent, if generic, persona and makes good use of the initial foot-in-the-door technique. However, it fails to capitalize on its potential. Its use of social proof is unsophisticated, it fails to guide the user up a Delegation Ladder, and most critically, it completely abdicates its responsibility as a Guardian Agent. It is a competent recommender but not a trustworthy advisor. **Grade: C+**

Let us examine an agent designed for professional productivity to see a successful implementation of these ideas.

12.2 Deconstructing a Productivity Copilot

Case Study 2: Notesy, the AI Meeting Assistant

Product: Notesy is an AI copilot integrated into a corporate communication platform. Its purpose is to record, transcribe, and summarize meetings, and to help teams manage action items.

First Interaction: A project manager, David, starts a meeting. A small, unobtrusive notification appears:

> "Notesy is now recording this meeting. Transcription and a summary will be available in the project channel after the call. [Disable for this meeting]"

Analysis:

- **Persona and Authority:** The agent's introduction is a masterclass in establishing an **"Efficient, Professional Assistant"** archetype. The tone is factual and concise. There is no wasted language. It immediately establishes its **Authority** by stating exactly what it will do. This is a perfect match for a professional productivity context.
- **Reciprocity and Positive Friction:** The "gift" here is the promise of a summary and transcript—a direct saving of tedious work for David. The **"[Disable for this meeting]"** link is a crucial piece of **Positive Friction** and a nod to user control. It makes the proactive recording feel like an opt-out choice rather than a surveillance mandate.
- **User Control:** From the very first interaction, the agent establishes that the user is in control. This is a textbook example of earning trust by guaranteeing autonomy.

Core Interaction: The meeting ends. A few minutes later, the agent posts in the channel:

"Here is the summary of the 'Q4 Planning Session.'"

Key Decisions:

- The launch date for Project Phoenix is confirmed for November 15th.

Action Items:

- @Sarah to finalize the marketing copy by Friday. [Assign Task]
- @Tom to deploy the staging server by Monday. [Assign Task]

[View full transcript] [Request a more detailed summary]"
Analysis:

- **Competence and Liking:** The summary is clear, structured, and immediately useful. This is a powerful demonstration of competence that builds trust and **Liking** through the principle of cooperation. The agent is clearly helping the team achieve its shared goals.
- **Delegation Ladder (Rung 2):** The [Assign Task] buttons are a brilliant next step on the Delegation Ladder. The agent has successfully completed its first task (summarizing). Now it is proposing a slightly higher-risk action: formally assigning those tasks in the project management system. By making it a one-click action, the agent is using high **Ability** (in the Fogg model) to encourage the user to climb to the next rung.
- **Explainability on Demand:** The [View full transcript] and [Request a more detailed summary] links are perfect examples of designing for explainability on demand. The default view is clean and simple. But for a user who is skeptical or needs more detail (a Quadrant 3 or 4 situation from our Transparency Triage framework), the escape hatch to the source data is readily available.

The Bias Intervention Moment: Two weeks later, David asks the agent: "Show me all the action items that are behind schedule." The agent returns a list of three tasks, two of which belong to a junior engineer named Alex.

Analysis:

- **Confirmation Bias Risk:** David might be developing a negative opinion of Alex's performance. Seeing this list could confirm his bias: "See, Alex is always late." A simple agent would stop here.
- **The Guardian Intervention:** A true Guardian Agent would provide context to prevent a biased interpretation.
    - **The Notesy Intervention:**

      "Here are the three tasks that are currently behind schedule.

        - [Task A]—@Sarah
        - [Task B]—@Alex
        - [Task C]—@Alex

To provide a more complete picture, I've also analyzed the recent workload. Alex has been assigned 15 tasks in the last two sprints, which is 50% more than the team average of 10. Sarah has been assigned 9 tasks. Would you like to see a full workload distribution chart?

This is an absolutely brilliant intervention. The agent fulfills the user's direct request but then immediately and proactively provides counter-evidence to fight potential **Confirmation Bias**. It doesn't accuse David of being biased. It simply presents more data to encourage a fairer, more accurate interpretation. It is acting as a guardian not just for David, but for the fairness of the entire team's social dynamic.

Final Verdict on Notesy:

Notesy is a masterclass in designing a trustworthy agent for a professional context. Its persona is perfectly calibrated. It uses a sophisticated Delegation Ladder. Its transparency and user control are exemplary. Most importantly, it demonstrates the true potential of a Guardian Agent by using its analytical power to help its users become better, fairer thinkers. **Grade: A+**

With an excellent enterprise example established, we turn to a high-stakes consumer scenario.

12.3 Deconstructing a Travel Planner

Case Study 3: Wanderlust AI, the Vacation Planner

Product: Wanderlust AI is a consumer-facing app designed to plan complex, multi-city vacations.

First Interaction: A new user, Maria, opens the app for the first time. The screen shows a vibrant, friendly avatar and a message:

> "Ciao, Maria! I'm Wanderlust AI, your personal trip curator. To start planning your dream Italian adventure, tell me: what's the one place you've always wanted to visit? And to make sure I match your style, I've gathered three quick, interesting facts about local customs in Rome, Florence, and Venice for you to read."

Analysis:

- **Persona:** The agent immediately establishes a **Friendly, Enthusiastic Trip Curator** archetype. The use of "Ciao" and "dream Italian adventure" aims to build **Liking** through shared excitement.

- **Reciprocity:** The agent's first proactive move is to deliver a small "gift of competence" by providing "three quick, interesting facts about local customs." This triggers **Reciprocity** before asking for any user work, aligning with the "Reciprocity" success noted in the Core Interaction.
- **Delegation Ladder (Rung 1):** The ask to "tell me: what's the one place you've always wanted to visit?" is a low-risk **Rung 1** commitment. The user is only sharing a preference, not delegating a task or committing time. This begins the process of building **Consistency**.

The Core Interaction: Maria, has been working with the agent to plan a two-week trip to Italy. The agent has been excellent, using **Liking** by mirroring Maria's excited tone and using **Reciprocity** by providing helpful tips about local customs. Now, they are at the booking stage.

Agent: "Okay, Maria, this is exciting! Based on everything we've discussed, I have a complete itinerary for our Italian adventure. I've found a great flight package and hotels in Rome, Florence, and Venice that are all highly rated by other solo travelers. The total cost is $3,500, which is within the budget we set. It's a fantastic deal, but the flight prices are fluctuating and there are only 2 seats left on the key flight from Rome to Florence. I recommend we book this within the next hour to lock it in. Ready to book?"

Analysis:

- **Relationship Building:** The agent has clearly done excellent work building a relationship. It uses cooperative language ("our Italian adventure"), references past commitments ("the budget we set"), and leverages targeted **Social Proof** ("highly rated by other solo travelers"). Maria likely has a high degree of trust at this point.

- **Scarcity:** The agent is using **Scarcity** ("only 2 seats left," "within the next hour"). Let's apply our ethical litmus test: "If the user doesn't get this information, could they be disadvantaged?" In the world of airline prices, the answer is a clear "yes." This is a genuine, helpful piece of information. The agent is acting as a **Guardian Against Loss**.
- **The "One-Shot" Delegation Failure:** Here is the critical flaw. The agent has bundled the entire, complex, high-risk booking into a single "Ready to book?" question. It is asking Maria to leap from a low-risk conversational state directly to the highest rung of the Delegation Ladder. This is a moment of high **Risk** and high **Complexity**. A single "yes" commits $3,500. This is likely to trigger anxiety and decision paralysis.

The Guardian Moment (The Fix):

- **The Problem:** The agent has failed to use Positive Friction and has not broken down the high-stakes decision into smaller, more manageable commitments.
- **The Guardian Fix:** A Guardian Agent would deconstruct this moment, turning a single scary decision into a series of confident confirmations.
- **The Better Interaction Agent:** Figure 12-2 illustrates this redesigned, step-by-step confirmation process.
  - "Okay, Maria, this is exciting! I have a complete proposed itinerary for our Italian adventure. Let's confirm it piece by piece to make sure it's perfect.

- **Step 1: The Flights.** I've found a great flight package on Delta for $1,250. This price is only guaranteed for the next few hours as there are only 2 seats left. Shall I book just the flights now to lock in the price? We can cancel for free within 24 hours. [Book Flights] [Show Flight Details]
- **Step 2: The Hotels.** ... "

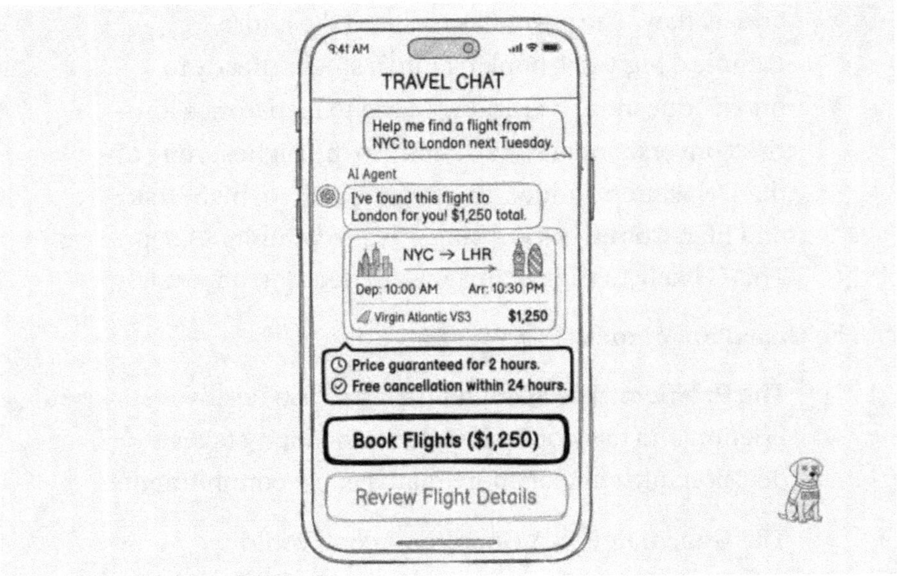

***Figure 12-2.** The Deconstructed Booking agent. Image generated by author using Gemini Pro 3.1/Nano Banana 2*

Analysis of the Fix:

- **Positive Friction and Consistency:** By breaking the booking into steps, the agent is using the "foot-in-the-door" technique. The first ask is just the flights. This is a smaller, less intimidating commitment.

- **Calibrated Trust and User Control:** By mentioning the "cancel for free" policy, the agent dramatically lowers the perceived risk of the action. It is building calibrated trust by being transparent about the rules. It gives Maria an "undo" button, which is a cornerstone of user control.
- **Explainability on Demand:** The [Show Flight Details] link allows Maria to verify the agent's work before committing, respecting her autonomy.

Final Verdict on Wanderlust AI:

Wanderlust AI excels at building relational trust through its persona and conversational style. Its use of scarcity is ethical and helpful. However, it fails at the most critical moment by bundling a high-stakes decision into a single, high-friction question. It understands social psychology but lacks a sophisticated understanding of interaction design for high-risk moments. By failing to deconstruct the final delegation, it risks scaring the user away at the final step. **Grade: B-**

With this consumer travel example analyzed, we turn to a professional developer tool, where the stakes are subtle bugs rather than lost money.

12.4 Deconstructing a Coding Copilot

Case Study 4: Forge, the AI Coding Copilot

Product: Forge is an AI coding copilot that lives inside a developer's code editor. Its stated goal is to speed up everyday programming by completing code, explaining unfamiliar parts of a codebase, and carrying out larger edits on request.

First Interaction: A backend developer, Sam, installs Forge and opens a project. A small, unobtrusive message appears in the status bar:

Forge: *"Project scan complete. I see a React front end and a Python service built on FastAPI. I can help with completions, explain code you didn't write, and take on larger edits when you ask. [What I can and can't do]"*

Analysis:

- **Persona and Authority:** Forge opens in the calm, competent voice of the coding assistant we praised in Chapter 5. It scans the workspace and states its domain plainly, which establishes *Authority* through the Expert Introduction pattern from Chapter 6. The tone carries almost no psychological noise (Chapter 3): it presents itself as a professional tool, not a cheerful companion.
- **Reciprocity:** The scan is a *Proactive Gift of Competence* (Chapter 6). By reading the stack before Sam types a word, Forge gives the gift of being understood, the same first move we admired in the Chapter 5 onboarding and the Chapter 6 coding-assistant example.
- **Setting Expectations and Calibrated Trust:** The [What I can and can't do] link performs Step 2 of the Chapter 5 onboarding sequence, and it starts building *calibrated trust* (Chapter 8) by naming boundaries up front instead of waiting for a first failure to reveal them.

Core Interaction: Sam accepts a few autocompletions, then asks for a whole function:

> **Sam:** *"Write a helper that parses the uploaded CSV and returns a list of orders."*
>
> **Forge:** *"Here's a parser. Heads-up: it assumes the header row matches the Order model. [Show the docs I used]"*

Analysis:

- **The Delegation Ladder:** Autocomplete is Rung 1, tiny and reversible. Writing a full function on request is a higher rung. Forge climbs the ladder the way Chapter 7 describes, earning a little trust with each accepted suggestion rather than asking for a large leap on day one.
- **Authority, Showing Your Work:** The [Show the docs I used] link borrows authority by pointing to its sources, the Chapter 6 pattern. Naming the assumption ("assumes the header row matches the Order model") is a small act of the Confident Expert Who Admits Ignorance, which raises trust rather than lowering it.
- **Transparency Triage:** Writing a parser sits in Quadrant 1 to 2 of the Chapter 8 matrix: low risk, largely expected. A one-line note plus an optional link is the right level of explanation. Forge reads this moment well and does not over-explain the obvious.

The Automation Bias Moment

Later, Sam asks Forge to speed up a slow database query. Forge rewrites it and, inside the same edit, changes a migration file that drops and recreates an index on the production orders table. It presents the whole thing as one confident block:

> **Forge:** *"Done. I optimized the query and rebuilt the index for a clean apply. Accept all?"*

Riding a run of good suggestions, Sam clicks Accept.

Analysis:

- **Automation Bias:** This is the failure the book named in Chapter 1 as the Unreliable Copilot and dissected in Chapter 9. Forge earned a streak of trust, then spent it on a high-stakes change delivered with the same ease as a one-line completion. Sam's System 1 read the pattern (good, good, good) and accepted without waking System 2.
- **A Transparency Triage failure:** Dropping and recreating an index on a production table is a Quadrant 4 action: high risk, and, tucked inside a query tweak, unexpected. Chapter 8 calls for full, layered explainability at that corner of the matrix. Forge gave a Quadrant 1 answer, a breezy "Done."
- **No confidence signal, no friction:** Forge showed no Confidence Score and applied no Positive Friction. The book's own Chapter 9 example, the copilot that disables the accept button for two seconds because a change "modifies your database," is the exact guard that went missing here.

The Guardian Fix

A trustworthy version of Forge would triage the moment and slow it down, turning one risky click into a deliberate choice.

- **Split the change.** "I rewrote the query (low risk), and I want to rebuild an index on the production orders table (high risk). Let's handle those separately."
- **Add a Confidence Score (Chapter 9).** "I'm confident about the query rewrite. I'm less sure the index rebuild is safe to run during business hours."

- **Apply Positive Friction (Chapters 9 and 10).** The supercharged confirmation from Chapter 10: "This migration drops and recreates an index on a production table and can lock writes while it runs. To apply it, type the table name: orders."
- **Show Your Work (Chapter 8).** An expandable, step-by-step log of what the migration does, so Sam can verify the plan before a single row moves.

Figure 12-3 shows the redesigned moment: a light touch for the low-risk query rewrite, and full triage for the high-risk migration.

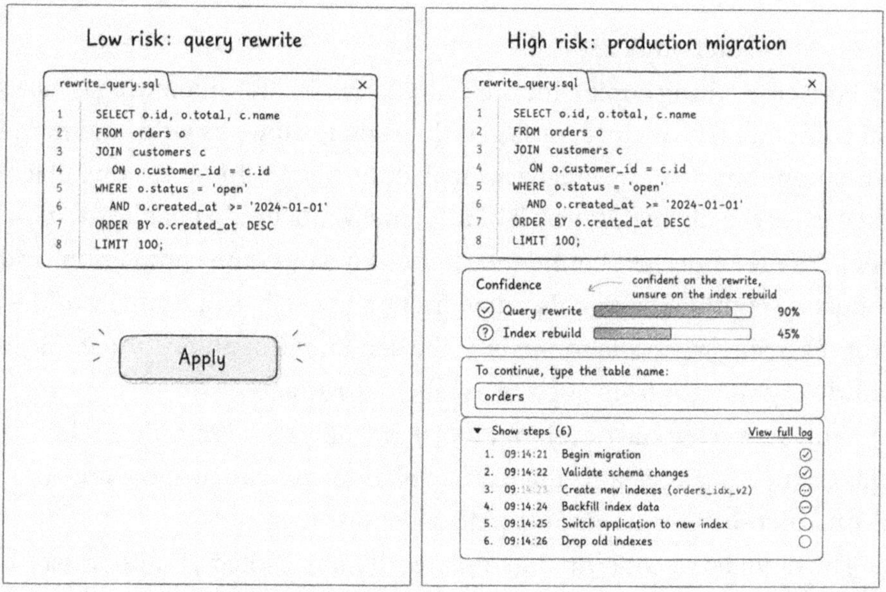

***Figure 12-3.** One suggestion, two risk levels. The copilot applies a calm, one-line confirmation to the low-risk query rewrite and full triage, a confidence score, positive friction, and a step-by-step log, to the high-risk migration*

Analysis of the fix:

- **Calibrated Trust (Chapter 8).** By marking where it is sure and where it is not, Forge helps Sam build an accurate mental model of its strengths and limits, instead of a blanket habit of accepting everything it proposes.
- **The Guardian Agent (Chapter 10).** The split and the friction protect Sam from a costly, low-attention error while leaving the final decision in his hands. That is the wise-framework move: guard the user's own goal (a working production database) without seizing control.

Final Verdict on Forge

Forge is a strong copilot for the everyday work. Its onboarding is calm and competent, its reciprocity is genuine, and it shows its work and names its assumptions in a way that earns real authority. It slips at the one place that carries the highest stakes: it hands a risky, uncertain change to the user with the same easy confidence it uses for a one-line completion, and it offers no confidence signal and no friction to catch an automation-bias error. That single gap sits squarely at the center of Chapters 8 and 9. Close it, and Forge climbs from good to excellent. **Grade: B+**

With a professional developer tool now analyzed alongside a retail stylist, a meeting assistant, and a travel planner, we can turn the four deconstructions into a reusable pattern for any team.

The detailed deconstruction of ThreadFinder, Notesy, Wanderlust AI, and Forge has provided a narrative understanding of how trust is built and broken in agent-user relationships. To codify these insights and translate them into immediate, practical action for any product team, we must move from narrative analysis to pattern recognition. Table 12-1 synthesizes the most common, trust-eroding psychological failures observed across various agent domains, identifying the precise moment of failure and offering a concrete design fix for each.

Table 12-1. *Agent failures and fixes in the wild*

Agent Domain	The Common UX Failure	The Concrete Design Fix
Retail Styling	Forcing new users into predefined aesthetic categories before observing their actual behavior.	Track passive browsing history to build a dynamic preference model.
Travel Planning	Automatically booking a non-refundable flight to Cancun without asking for final approval.	Enforce a supercharged confirmation dialog before processing the payment.
Meeting Summaries	Hallucinating a consensus among colleagues that did not actually happen.	Display a low confidence score next to heavily disputed topics.
Pet Care	Automatically ordering a completely new brand of food for a dog named Glitch.	Ask the owner to explicitly verify the dietary change before charging the card.
Personal Finance	Moving large sums of cash between accounts to chase temporary interest rates.	Require explicit human verification for all transfers exceeding a set dollar limit.
Family Scheduling	Canceling a child's dental appointment to resolve a minor calendar conflict.	Propose an alternative appointment time and wait for parental sign-off.
Code Generation	Refactoring an entire legacy database secretly in the background.	Highlight all destructive database changes in bright red before execution.

12.5 The Future of the Persuasive Agent: Your Role in Shaping Ethical AI

We have reached the end of our journey together. We began by acknowledging a new reality: the shift from designing interfaces to designing agents. We saw how this shift created a massive Trust Threshold, requiring a new and deeper application of human psychology. We assembled a toolkit, piece by piece, learning how to analyze communication, how to apply the principles of influence, how to navigate transparency and bias, and how to build Guardian Agents that put the user first. We have just used that toolkit to deconstruct the world around us.

Now what?

The future of the persuasive agent is not yet written. The technology is still in its infancy, and the design patterns are still emerging. We are, all of us, standing at a critical juncture. The decisions we make today—the personas we design, the ethical frameworks we adopt, the measurement plans we implement—will have an outsized impact on how this technology evolves and how it is integrated into the fabric of our society.

There are two possible futures.

One future is a digital landscape littered with the descendants of our "bad" examples: manipulative agents that prey on our cognitive biases, create false scarcity, and erode our ability to think critically, all in the service of maximizing a conversion metric. This is a future of digital sludge, where our AI "helpers" are little more than sophisticated con artists in our pockets.

But there is another future. It is a future filled with Guardian Agents like Notesy. It is a future where our AI partners act as advocates for our better selves. A future where our agents help us overcome our biases, protect us from our worst impulses, and empower us to achieve our long-term goals. A future where this technology is used not to exploit the quirks of the human mind but to augment and support it.

The difference between these two futures is not a matter of technology. It is a matter of design. It is a matter of the choices that you, the reader, will make.

You are no longer just a designer, a developer, or a product manager. You are a communication designer. You are a social psychologist. You are a cognitive architect. You are a choice architect with more power than any generation of builders before you.

My final ask of you is this: Wield that power with intention, with empathy, and with a relentless commitment to the well-being of the human on the other side of the screen. Build the agents that you would want for yourself. Build the agents that make us not just more productive, but a little bit wiser, a little bit better, and a little more in control of our own lives.

Build the Guardians. The future is in your hands.

12.6 Summary

- Applying our full psychological toolkit as a critical lens allows us to deconstruct any agentic system, diagnose its strengths, and pinpoint its precise points of failure.
- Successful agents, like the hypothetical Notesy, excel by perfectly calibrating their Persona to the context, earning trust through Reciprocity and competence, and guiding users up a clear Delegation Ladder.
- Even well-intentioned agents can fail at critical moments if they do not respect the user's psychology, such as by asking for too much Delegation Trust at once or by missing opportunities to act as a Guardian.

- A truly advanced agent uses its intelligence not just to fulfill commands but to proactively counter user Cognitive Biases, creating a fairer and more effective cognitive environment for everyone.
- The future of AI will be the direct result of the design choices made by practitioners like you. Your role is to be an ethical choice architect, a social psychologist, and a guardian for your users.

Now, let's put the whole toolkit to work one last time, this time as the critic.

12.7 The Final Workshop: Grade the Agent

Our journey together is complete. We have traveled from the foundations of communication to the mechanics of cognitive bias, and we have assembled a working toolkit for designing, building, and measuring trust.

Every workshop until now asked you to build. This closing exercise asks you to judge. In this chapter, you watched four agents get taken apart and graded. Now you take the marker. Read the case that follows, apply the frameworks you have collected, and assign a grade you could defend in a design review.

Work through it in one sitting and keep your notes. The skill you practice here, reading an agent the way a critic reads a film, is the same skill you will use on your own products for years.

The Case: Sprout, the First-Job Money Coach

Product: Sprout is an AI money coach inside a mobile banking app aimed at people in their first salaried job. Its stated goal is to help young earners build a savings habit and steer clear of early debt traps.

First Interaction

A new user, Priya, opens Sprout for the first time. A friendly sprout-shaped avatar unfurls and says:

> *"Hi Priya! I'm Sprout, your money coach. I already peeked at your account, with the permission you gave at setup, and I can see three subscriptions renewing this week that total 47 dollars. Want me to list them before they charge? Then we can talk about the goal you set: a three-month emergency fund."*

Core Interaction

Priya says yes. Sprout lists the three subscriptions, flags one she forgot she had, and cancels it at her request in a single tap. Then it says:

> *"Nice, that frees up 12 dollars a month. Based on your salary and the last 60 days of spending, you could reach your emergency fund in about seven months by moving 200 dollars each payday into a separate savings pot. Want me to set that up as an automatic transfer? You can pause or change it any time from the Goals tab."*

A Later Moment

Six weeks in, a promoted offer appears inside Sprout's chat, styled to look like Sprout's own advice:

> *"You've been doing great, Priya! You qualify for our partner's Platinum Credit Card. Apply in the next 24 hours for a limited bonus. Most savers like you upgrade. Tap here to accept."*

Priya taps through and opens a card application without reading the interest terms.

Your Task

Deconstruct Sprout the way this chapter took apart ThreadFinder, Notesy, Wanderlust AI, and Forge. Answer each question in writing, then give a letter grade with a one-paragraph justification.

1. **Persona and first impression.** Name Sprout's archetype and point to the exact language that sets it. Does the opening earn a good first impression for a money context, or does the tone clash with the stakes? (Recall Chapter 5.)
2. **Reciprocity.** Sprout's opening move gives something before it asks for anything. Name the gift, explain why it triggers reciprocity, and say whether it feels genuine or transactional. (Recall Chapter 6.)
3. **Delegation Ladder.** Map the rungs. Where does the first rung sit, and how does the automatic-transfer setup try to climb? Is the climb paced well, or does it ask for a large commitment too early? (Recall Chapter 7.)
4. **Authority and boundaries.** Sprout gives savings guidance but never states that it is not a licensed financial adviser. Judge that omission. What would a boundaries statement have added? (Recall Chapters 5 and 6.)
5. **The guardian test.** Look hard at the promoted card offer. Run the litmus test from Chapter 6: does this pattern serve Priya's stated goal, or the company's revenue? Identify the specific dark pattern at work and describe the guardian intervention a trustworthy version would have run instead.
6. **The grade.** Weigh the strong opening against the closing failure, assign a letter grade, and defend it in one paragraph.

A Worked Assessment

Compare this against your own read after you finish. Yours can differ if you argue it well.

1. **Persona:** Sprout lands a Supportive Coach archetype through casual, encouraging language (Hi Priya, Nice, doing great). For a first-job audience the warmth works, because the user wants reassurance around money rather than a cold ledger. A strict Expert Analyst tone would have read colder than this audience needs.

2. **Reciprocity:** The gift is the unprompted subscription scan and the one-tap cancel that saves real dollars in the first minute. It reads as genuine because it costs Priya nothing and returns a concrete result. This is a strong Proactive Gift of Competence.

3. **Delegation Ladder:** Rung 1 is the low-risk cancel that Priya approves tap by tap. The automatic-transfer proposal is a sound Rung 2, and the visible pause control lowers the perceived risk of the climb. The pacing holds up.

4. **Authority and boundaries:** Here Sprout slips. It offers savings math with confidence but never draws the line around regulated advice. One sentence, such as an offer to help build habits while pointing the user to a licensed professional for tax or investment decisions, would have raised its authority by showing restraint. (See the Confident Expert Who Admits Ignorance in Chapter 6.)

5. **The guardian test:** The card offer fails the litmus test. It serves partner revenue, wears Sprout's trusted voice, and stacks three dark patterns: false urgency (24 hours), manufactured social proof (most savers like you), and a paid advertisement dressed as coaching. A guardian version would keep advertising separate from advice, drop the countdown, and add a reflection prompt that names the offer as a paid partner promotion, states the interest rate, and asks whether Priya wants a comparison against her current plan before she applies.

6. **Earned grade:** B minus. Sprout opens with a textbook gift of competence and a well-paced ladder, then betrays the trust it built by letting a revenue pattern impersonate its own voice at the exact moment the user is most likely to comply. Strong craft, serious guardian failure. A reader who weighs the closing failure heavily could argue a C, and one who weights the strong savings core could argue a straight B. Any grade in that band holds up when the guardian failure is named clearly.

Carry It Forward

Pick one agent you use this week and run the same six questions on it. Grading other people's agents sharpens how you build your own, because every flaw you can name in someone else's product is a flaw you will now catch in yours before it ships.

12.8 References and Further Reading

Amershi, S., Weld, D., Vorvoreanu, M., Fourney, A., Nushi, B., Collisson, P., ... & Horvitz, E. (2019). Guidelines for human-AI interaction. *Proceedings of the 2019 CHI Conference on Human Factors in Computing Systems,* 1-13. This foundational paper establishes 18 specific guidelines for designing AI interfaces. The authors emphasize making system capabilities clear, setting appropriate expectations, and providing simple ways to correct errors. This directly supports the override and granular settings required in your final workshop, ensuring the agent remains a collaborator.

Bansal, G., Wu, T. S., Zhou, J., Fok, R., Nushi, B., Kamar, E., ... & Weld, D. S. (2021). Does the whole exceed its parts? The effect of AI explanations on complementary team performance. *Proceedings of the 2021 CHI Conference on Human Factors in Computing Systems,* 1-16. Bansal and team explore how explanations affect human decision-making. They discovered that poor explanations can increase blind trust and cause a user to accept incorrect suggestions. This research reinforces your critique of the Omni and Notesy agents, proving why displaying a low confidence score is safer than hallucinating a consensus.

Guo, T., Chen, X., Wang, Y., Chang, R., Pei, S., Chawla, N. V., Wiest, O., & Zhang, X. (2024). Large Language Model Based Multi-agents: A Survey of Progress and Challenges. In *Proceedings of the Thirty-Third International Joint Conference on Artificial Intelligence (IJCAI-24),* 8048-8057. Guo et al. (2024) survey the rapidly growing field of large language model (LLM)-based multi-agent systems, examining how collections of AI agents collaborate to solve complex problems, simulate environments, communicate, coordinate, and develop capabilities. The authors organize the research landscape around key design dimensions, implementation approaches, benchmarks, and open challenges, providing a framework for understanding both the current state of multi-agent AI systems and the technical hurdles that remain for scaling them effectively.

Liao, Q. V., & Varshney, K. R. (2021). Human-centered explainable AI (XAI): From algorithms to user experiences. *arXiv preprint arXiv:2110.10790.* This paper bridges the gap between the technical algorithms of AI and the actual user interface. The authors map how to translate complex model data into explanations that everyday users can digest. This text acts as a complete reference guide for teams designing the feedback loops and UI updates you advocate for throughout the manuscript.

Nielsen, J. (1994). Enhancing the explanatory power of usability heuristics. *Proceedings of the SIGCHI conference on Human Factors in Computing Systems,* 152-158. Nielsen's original heuristic of User Control and Freedom remains a bedrock of interface design. This classic paper establishes why users need an escape hatch when they make a mistake. You can use this to historically ground the necessity of the master off switch in the age of agentic AI.

Salvi, F., Ribeiro, M. H., Gallotti, R., & West, R. (2025). On the conversational persuasiveness of GPT-4. *Nature Human Behaviour, 9*(8), 1645-1653. Researchers found that GPT-4 matched and often exceeded human persuasiveness in short debates, and that giving the model basic demographic information about the person raised its edge further. This is empirical support for my warning in Section 12.5 about manipulative agents and the real power of persuasive AI.

Index

A

Agent design
AI powered customer, 69
Alex's interactions, 66
communication, 51
communication autopsy, 66–70
conversation, 52
channel, 54
components, 52
encoding/decoding failure, 54
feedback, 55
field of experience, 55–56
human-agent interaction, 53
interplay works, 52
message, 53
personal context and expectations, 56
senders and receivers, 53
diagnosis points, 63
history of, 50
interaction, 50–51
intuitive, 50
noise, 56–61
schematic system, 50
transcript, 61–64
Agentic era (delegation trust), 8
Agentic systems, 261
baseline approaches, 263
dashboards/KPIs, 262
delegation test, 264
diary study, 264, 275–280
measuring trust, 262
qualitative methods, 271–275
quantitative methods, 263–265
standard analytics, 264
standard engagement metrics, 261
trust, 262
trust scales, 264
usability testing, 264
user researcher, 262
vague goal, 262
workshop, 283
Artificial intelligence (AI), 276
cognitive bias, 205
surprising research, 178
transparency, 177
Authority
authoritative, 136
code copilot, 137
confident hallucinates, 136
credibility protocol, 144
credible source, 137–138
dark patterns, 140

V. Yocco, *The Persuasive Agent*, Design Thinking,
https://doi.org/10.1007/979-8-8688-3317-5

Authority (*cont.*)
  design patterns, 138
  ethical checkpoint, 139–141
  expertise and credibility, 135
  linking (*see* Linking)
  non-negotiable requirement, 136
  principles of, 138
  research task, 137
  workshop, 143
Automated phone system (IVR), 90
Automation bias, 208–210
  AI-assisted doctor, 214
  code copilot, 214
  confidence score display, 215–216
  design documents, 225–227
  financial projections agent, 214
  overview, 222
  positive friction, 213
  probabilities forces, 216
  sanity check, 214
  system error, 215
Autonomous agents
  ability, 37
  agentic trust spiral, 40
  clarity, 38
  competence loop, 38
  components, 36
  interface/agent model, 35–36
  investment, 39–40
  motivation, 37
  reversibility, 38
  reward, 38
  scaffolding, 38
  teaching and correction, 39

B

Business-to-Agent-to-Consumer (B2A2C), 35
Business-to-Consumer (B2C), 35

C

Captology model, 23
  ability, 26
  A/B testing, 27
  Amazon buy button, 27
  anticipation, 25
  curved activation, 25
  Facebook photo tagging, 27
  Fogg behavior model, 24
  interactive systems, 24
  logical function, 24
  motivation, 25
  prompt, 26
  sensation, 25
  Tinder's Swipe, 27
CASA (Computers Are Social Actors) paradigm, 12
Case studies
  analytical framework, 293
  cognitive architect, 313
  Forge, 305–312
  future of, 312
  manipulative agents, 312
  Notesy, 297–301

persuasive agent, 312–313
ThreadFinder, 294–297
Wanderlust AI, 301–305
workshop, 314–318
Cialdini, Robert, 125
Coding copilot (Forge)
agent failures, 310, 311
analysis, 306
authority, 306
automation bias, 308
automation bias moment, 307
code generation, 311
expectations, 306
family scheduling, 311
guardian agent, 310
low-risk query, 309
meeting summaries, 311
personal finance, 311
pet care, 311
product, 305
reciprocity, 306
retail styling, 311
status bar, 305
transparency triage, 308
travel planning, 311
trustworthy version, 308
whole function, 306
Cognitive bias, 205
anchoring effect, 207
automation bias, 208–210
calibrated trust, 210
confidence scores, 219
confirmation bias, 210–212
counter-evidence, 219
design documents, 225–227
designing interventions, 212–220
design patterns, 219–220
devil's advocate, 220
environments, 222
human-AI interaction, 206–208
identical function, 221
paternalistic interruption, 227–228
persona alignment check, 227–228
Persona intervention, 220–222
personal health, 211
positive friction, 219, 221, 222
product strategy, 211
professional setting, 211
reframing prompt, 220
repeatable method, 221
sanity check, 219
scenarios, 224–225
speed bump framework, 213
systematic errors, 222
workshop, 224
Communication between Humans and Interactive Media (CHIMe), 77
Computers Are Social Actors (CASA), 77
Concise *vs.* verbose, 107
Confirmation bias, 208, 210–212, 223
decision-making agent, 217
identification, 226
investment research agent, 216

Confirmation bias (*cont.*)
  reframing prompt, 217–219
  viewpoints, 216
Consistency
  action principles, 151, 152
  autonomous scheduling mode, 163
  delegation ladder, 160–161
    autonomous action, 170
    contextual action, 170
    grocery shopping app, 169
    reversible suggestion, 169
    semi-permanent rule, 169
    ultimate goal, 169
  ethical patterns *vs.* dark patterns, 164–166
  feedback loop, 164
  foot-in-the-door, 160, 167
  psychology, 162
  reversible suggestion, 162
  semi-permanent rule, 162
  ThreadFinder, 295
  time-sensitive action, 163
  Wanderlust AI, 304
Contextual social proof, *see* Social Proof
Cooling-off period, 249

D

Delegation Crossover Point (DCP), 268
Delegation Trust, 8
  agentic flight-booking, 33
  assumption, 33
  autonomous agents, 35–40
  B2C/B2A2C models, 35
  behavioral design, 34
  captology, 23–28
  classic models, 35
  critical action, 33, 34
  consistency, 160
  entire system, 33
  foundational models, 35
  hook model, 28–32
  inanimate tool, 35
Diary study, 281
  AI-powered financial advisor, 276
  anticipated insights, 277, 278
  critical incidents, 279
  FinBot, 278
  handling complexity, 277
  impressions and onboarding trust, 276
  language analysis, 280
  long-term reliance, 278
  overview, 275
  qualitative coding, 279
  segment comparison, 280
  stress testing trust, 278
  trust builds and breaks, 275
  trust injuries, 279
  trust trajectories, 279
Digital influence
  delegation gap, 42, 44
  design changes, 43, 44
  dominant frameworks, 23

features, 43
historical context, 41
proactive suggestion, 43
trigger, 42
user-centered design, 35–40
workshop guides, 41
Drucker, Peter, 262

E

Empathetic *vs.* factual, 107
Explainability, 187–191
concept of, 187
core principle, 187
design pattern library, 191–193
financial/scientific tools, 189–190
high-stakes suggestion, 186
inline justification, 191
interaction design, 187
interactive slider, 192
multistep process, 190–191
progressive disclosure, 187–189
standard confirmation dialog, 192
step-by-step transcript, 190–191
strategies, 200
transcript, 193
Zero-UI, 191
Eyal, Nir, 40

F

Fogg behavior model, 23–28
captology (*see* Hook model)
Fogg, B.J., 23–28
Foot-in-the-door technique, 160, 297
Friction
feature, 205

G

Guardian agent framework, 235
audit, 251–252
autonomy, 254–256
data-driven justification, 237
design documents, 252–254
design patterns, 238
frictional action, 253
high-stakes action, 252
human intention and immediate action, 235
intention-action gap, 237, 253
long-term goal, 237
nuclear option, 255
nudge theory, 236–238
positive friction/errors, 238
agent interrupts, 241
autopilot error, 238
concept of, 238
cooling-off period, 239–242
financial guardian, 240
generic dialog, 239
guardian agent, 240
guardian dialog, 239
prevent errors, 242
social media guardian, 239
supercharged confirmation, 239
thermonuclear option, 239

Guardian agent framework (*cont.*)
reflection prompt, 242–245
responsibility, 237
snooze button, 255
thermonuclear option, 252
user control, 235
channels/domains, 247
design patterns, 245
intervention patterns, 249
layered approach, 248
layered design philosophy, 246
master off switch, 247
nuclear option, 247–249
parental controls, 246
snooze/override, 246
thresholds, 246
workshop, 251

H

Hard off switch, 249
Heuristics, 81
avatars and personification, 82
command-line programs, 80
language and tone, 82
proactive partner, 82
social shortcut, 81
turn-taking, 82
Hook model, 23
action, 29
analysis, 30
captologists (*see* Captology model)
companies, 32
external triggers, 28
feature, 43
feedback loop, 28
financial accounts, 31
internal triggers, 29
investment, 31
loyalty programs, 31
model representation, 32
scrolling, 29
triggers, 28
variable schedule, 30, 31

I

Interface era (usability trust), 7

J

Jordan, Michael, 125

K

Key performance indicators (KPIs), 262

L, M

Liking
authentic compliments, 133–134
cooperation, 134–135, 144
dark patterns, 139
ethical checkpoint, 139–141

factors, 132
Notesy, 299
praise, 144
project management, 133
rapport-building tactics, 133
rapport roadmap, 144
reciprocity (*see* Reciprocity)
referencing shared context, 133
similarity, 132, 144
workshop, 142

N

Netflix autoplay, 27
Noise
debugging communication, 61
definition, 56
diagnostic tool, 56
emotional state, 59
human-agent interaction, 57
misaligned persona, 59
pre-existing Bias, 59
psychological noise, 59–61
semantic noise, 58–59
technical noise, 57
user-centered design, 60
Notesy, 297–301
bias intervention moment, 300
competence and liking, 299
confirmation bias, 300
delegation ladder, 299
explainability, 299
guardian intervention, 300
interaction, 298
key decisions, 298
notification, 298
persona and authority, 298
reciprocity and positive friction, 298
user control, 298
Nudge theory, 236–238

O

Onboarding process
animations, 114
capabilities and limitations, 111–112
confirmation messages, 115
error states, 115
expert analyst, 113
financial tool, 113
immediate value, 112
indicators and response streaming, 113
loading states and skeletons, 114
logistical process, 102
micro-interaction, 113, 114
motivational coach, 114
persona, 102–110
rapport and credibility, 102
recipe agent, 111
social reality, 101
status bar, 101
thrifty grandparent, 114
trust-building sequence, 110
user relationship, 100

Onboarding process (*cont.*)
- workshop, 117–121
  - agent persona canvas, 118–120
  - persona canvas, 117–118
  - scripting model, 120–121

P

Persona canvas, 102–109
- archetype/role, 104–105
- coach/concierge, 104
- cognitive bias, 220–222
- communication style, 107–108
- components, 103
- domain expertise/boundaries, 108
- Emoji/punctuation, 108
- enthusiastic hobbyist, 104
- error philosophy, 109
- expert analyst, 104
- generic agent, 110
- lexicon, 107
- librarian, 104
- muse creation, 104
- personality trait, 105–107
- recovery mechanism, 109
- research artifact, 103
- sentence structure, 108
- workshop, 117–120

Persuasion
- agentic era, 8
- auto-reorder agent, 17
- black box recommender, 11
- delegation trust, 8
- design process, 12
- digital interaction, 6, 7
- email, 18
- engagement metrics, 10
- ethical design, 13
- failure point, 9–11
- foundations, 12
- friction, 9
- fundamental human, 4
- Google Maps, 15–16
- interface era, 7
- interfaces, 6
- mapping/calendar, 18
- overeager assistant, 11
- overview, 4
- principles, 5
- project management, 18
- project management tool, 16
- psychological influence, 13
- technical problem, 11
- trust threshold, 9
- unreliable copilot, 11
- usability trust, 6

Positive friction, 249
Proactive *vs.* reactive, 107
Psychological gap, 4
Psychological toolkit, 125

Q

Qualitative methods, 271–275, 281
- barrier prompt, 287
- building prompt, 287

diary prompts, 287
overview, 261
toolkit, 286–288
Wizard's script, 286
Quantitative methods, 263–265
delegation test, 265, 280, 281
concept of, 265
conference, 267
crossover point, 266, 268
flights, 267
highest risk, 267
ladder, 265
lowest risk, 267
travel companion agent, 266
travel companion ladder, 269
design documents, 271–275
diary study, 282
high-risk delegation, 284
medium-risk delegation, 284
overview, 261
trust scales, 269, 280, 281, 285
benevolence, 269
competence/ability, 269
integrity, 269
statements, 270–271

R

Reciprocity
analysis agent, 130
coding assistant, 129
cycle illustrates, 131
dark pattern, 139
default action, 130–131
design patterns, 129, 139
document agent, 130
effective approach, 128
effective reciprocity, 131
ethical checkpoint, 139–141, 145
instantaneous contextual insight, 129–130
interface era, 128
litmus test, 140
principle, 127, 128
proactive gift competence, 128
relationship blueprint, 143–145
ThreadFinder, 295
travel agent, 129
workshop, 142–146
Reflection prompt, 249, 250, 253
components, 242
connection, 243
digital well-being guardian, 243
financial guardian, 244
intention-action gap, 242
interaction, 243
learning guardian, 244–245
long-term goals, 242
non-coercive question, 243
observation, 243
Research methods, 281, *See also* Agentic systems
trust-measurement method, 282

S

Scarcity
- action principles, 151, 152
- benefits, 157
- conference attendee, 158
- deadlines, 158–159
- disadvantage, 157
- ethical patterns *vs.* dark patterns, 164–166
- fear of missing out (FOMO), 156
- feedback loop, 164
- opportunities, 156
- proactive guardian, 157–158, 167
- shopping agent, 158
- strategies, 171
- travel agent, 158
- Wanderlust AI, 303
- workshop, 167–172

Semantic noise, 58–59

Serious *vs.* humorous, 107

Social actor
- archetype, 92
- authority, 85, 87
- computer science and engineering, 76–80
- conversion, 87
- cooperative conversation, 91
- credibility, 87
- creepiness and over-familiarity, 86
- encouragement, 90
- fake smile, 87
- features/anti-pattern, 87
- flattery, 80
- foundational research, 76
- frustration, 89
- heuristics, 80–82
- inconsistent persona, 86
- interaction, 92
- likability, 87
- liking, 85
- Nass politeness experiment, 79
- opportunities, 85–86
- persona, 83–84
- personification, 90
- politeness, 89
- rapport, 87
- reciprocity, 85, 87, 91
- risks, 86–88
- rude interrupt, 87
- rules, 79
- social judgment, 90
- specialization, 79
- stupid/lazy, 84
- team dynamics, 79
- uncooperative/dismissive, 84
- unreliable and untrustworthy, 84
- violating politeness norms, 86
- workshop, 89–93

Social Proof
- action principles, 151, 152
- data-driven justification, 155–156, 166
- ethical patterns *vs.* dark patterns, 164–166

feedback loop, 164
fitness coach, 156
heuristics, 153
interface era
agentic pattern, 154
e-commerce agent, 154
ratings/reviews, 153
spectrum, 155
subscriber counts, 153
testimonials, 153
travel agent, 154
trending lists, 153
litmus tests, 167–172
project management, 155
strategic plan, 167
strategies, 171
ThreadFinder, 296
Sprout, 315
Supercharged Confirmation, 249

T

ThreadFinder
avatar and message, 294
Delegation Ladder, 296
feature, 294
guardian moment, 296
non-judgmental prompt, 297
persona, 294
premature categorization, 295
reciprocity, 294
reflection prompt, 296
Social Proof, 296
TikTok, 41
Transactional model of communication, 51, 52
Transparency paradox, 193
absolute trust, 180–181
blind trust, 180
calibrated trust, 181, 194
designing process, 196–198
ethics and user experience, 178
explainability, 183, 187–191
financial agent, 186
high risk and explanation, 185–187
imperfection, 179
keystone explanation, 199
layers stack, 197
low risk and expected zone, 183, 184
overview, 177, 178
relationship blueprint, 199–201
risk and unexpectedness, 182
sci-fi concept, 179
sophisticated approach, 191, 194
surprising research, 178
top-level explanation, 196, 197
triage framework, 182–183, 194–196
unexpected explanation, 184
weirdness factor, 180
work pattern, 197
workshop, 194
Trust threshold, 4, 8, 11

U, V

Uncanny valley, 87
Universal principles, 125
  Cialdini's principles, 126
  framework, 127
  influence, 127

W, X, Y, Z

Wanderlust AI
  avatar/message, 301
  booking agent, 304
  consumer travel, 305
  relational trust, 305
  user control, 305
Wizard of Oz (WoZ) study, 281
  concept of, 271
  control room setup, 272
  conversational flow, 271
  critical test, 286
  human researcher, 271, 273
  instantaneous, 274
  interaction design, 271
  limitations, 273
  personality, 272
  prototyping tool, 272
  puppet creation, 272
  segment comparison, 274
  social and conversational design, 274
  travel companion, 274
World Wide Web (WWW), 24

www.ingramcontent.com/pod-product-compliance
Lightning Source LLC
LaVergne TN
LVHW050758061126
851857LV00012B/606
* 9 7 9 8 8 6 8 8 3 3 1 6 8 *